Further Electrical and Electronic Principles

Further Electrical and Electronic Principles

Third edition

Christopher R Robertson

Routledge
Taylor & Francis Group

LONDON AND NEW YORK

First published 1993 by Newnes as *Electrical and Electronic Principles 2*
by Edward Arnold
Second edition 2001
Third edition 2008

Published 2015 by Routledge
2 Park Square, Milton Park, Abingdon, Oxon OX14 4RN
711 Third Avenue, New York, NY 10017, USA

First issued in hardback 2017

Routledge is an imprint of the Taylor & Francis Group, an informa business

British Library Cataloguing-in-Publication Data
A catalogue record for this book is available from the British Library

Library of Congress Cataloging-in-Publication Data
A catalog record for this book is available from the Library of Congress

ISBN 13: 978-1-138-41339-9 (hbk)
ISBN 13: 978-0-7506-8747-8 (pbk)

Typeset by Charon Tec Ltd (A Macmillan Company), Chennai, India.
www.charontec.com

Contents

Preface

This textbook supersedes the Second Edition of the same title. In response to feedback from colleges, the material has been modified in order to closely reflect the syllabus content of the BTEC Unit Further Electrical Principles. At the end of the chapters in the previous edition there were extra worked examples in the form of Supplementary Worked Examples. The majority of these have now been incorporated within the text of the relevant chapters, and those which have not may be accessed online at the website address www.routledge.com/ 9780750687478, together with removed chapters on Bipolar and Field Effect Transistors, Transistor Circuits, Analysis of Small-signal Transistor Amplifiers, Operational Amplifiers, The Decibel and its Usage, and Control Principles. These chapters may be of use to students wishing to follow electrical/electronic pathways for further studies.

Further Electrical and Electronic Principles contains 327 illustrations, 85 worked examples, 17 suggested practical assignments, and 111 assignment questions, the answers to which are to be found towards the end of the book.

This book continues with the philosophy of the original in that it may be used as a complete set of course notes for students studying the unit *Further Electrical Principles* as part of the second year of a BTEC National Diploma/Certificate course, or any other course that incorporates this unit. Coverage of the BTEC Unit *Electrical and Electronic Principles* is contained in the third edition of the companion book, *Fundamental Electrical and Electronic Principles*, ISBN 9780750687379.

C. Robertson
Tonbridge
March 2008

Preface

This textbook supersedes the Second Edition of the same title. In response to feedback from colleges, the material has been modified in order to closely reflect the syllabus content of the BTEC Unit Further Electrical Principles. At the end of the chapters in the previous edition there were extra worked examples in the form of Supplementary Worked Examples. The majority of these have now been incorporated within the text of the relevant chapters, and those which have not may be accessed online at the website address http://books.elsevier.com/companions/9780750687478, together with removed chapters on Bipolar and Field Effect Transistors, Transistor Circuits, Analysis of Small-signal Transistor Amplifiers, Operational Amplifiers, The Decibel and its Usage, and Control Principles. These chapters may be of use to students wishing to follow electrical/electronic pathways for further studies.

Further Electrical and Electronic Principles contains 327 illustrations, 85 worked examples, 17 suggested practical assignments, and 111 assignment questions, the answers to which are to be found towards the end of the book.

This book continues with the philosophy of the original in that it may be used as a complete set of course notes for students studying the unit *Further Electrical Principles* as part of the second year of a BTEC National Diploma/Certificate course, or any other course that incorporates this unit. Coverage of the BTEC Unit *Electrical and Electronic Principles* is contained in the third edition of the companion book, *Fundamental Electrical and Electronic Principles*, ISBN 9780750687379.

C. Robertson
Tonbridge
March 2008

Chapter 1

Single-Phase Series A.C. Circuits

Learning Outcomes

This chapter concerns the effect of resistors, inductors and capacitors when connected to an a.c. supply. It also deals with the methods used to analyse simple series a.c. circuits. At the end of the chapter, the concept of series resonance is introduced.

On completion of this chapter, you should be able to:

1 Draw the relevant phasor diagrams and waveform diagrams of voltage and current, for pure resistance, inductance and capacitance.
2 Understand and use the concepts of reactance and impedance to analyse simple a.c. series circuits.
3 Derive and use impedance and power triangles.
4 Calculate the power dissipation of an a.c. circuit, and understand the concept of power factor.
5 Explain the effect of series resonance, and its implications for practical circuits.

1.1 Pure Resistance

A pure resistor is one which exhibits only electrical resistance. This means that it has no inductance or capacitance. In practice, a carbon or metal film resistor is virtually perfect in these respects. Large wire-wound resistors can have a certain inductive and capacitive effect.

Consider a perfect (pure) resistor, connected to an a.c. supply, as shown in Fig. 1.1. The current flowing at any instant is directly proportional to the instantaneous applied voltage, and inversely proportional to the resistance value. The voltage is varying sinusoidally, and the resistance is a constant value. Thus the current flow will also be sinusoidal,

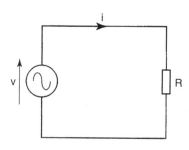

Fig. 1.1

and will be in phase with the applied voltage. This can be written as follows

$$i = \frac{v}{R} \text{ amp}$$

$$\text{but, } v = V_m \sin \omega t \text{ volt}$$

$$\text{therefore, } i = \frac{V_m}{R} \sin \omega t \text{ amp}$$

$$\text{but, } i = I_m \sin \omega t \text{ amp}$$

Thus, the current is a sinewave, of maximum value V_m/R, is of the same frequency as the voltage, and is in phase with it.

Hence,

$$I_m = \frac{V_m}{R} \text{ amp, or } I = \frac{V}{R} \text{ amp} \tag{1.1}$$

The relevant waveform and phasor diagrams are shown in Figs. 1.2 and 1.3 respectively.

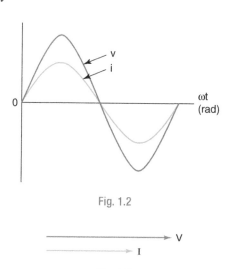

Fig. 1.2

Fig. 1.3

The instantaneous power (p) is given by the product of the instantaneous values of voltage and current. Thus $p = vi$. The waveform diagram is shown in Fig. 1.4. From this diagram, it is obvious that the power reaches its maximum and minimum values at the same time as both voltage and current. Therefore

$$P_m = V_m I_m$$

$$\text{hence, } P = VI = I^2 R = \frac{V^2}{R} \text{ watt} \qquad (1.2)$$

Note: When calculating the power, the *r.m.s.* values must be used.

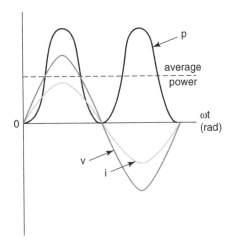

Fig. 1.4

From these results, we can conclude that a pure resistor, in an a.c. circuit, behaves in exactly the same way as in the equivalent d.c. circuit.

Worked Example 1.1

Q　Calculate the power dissipated by a 560 Ω resistor, when connected to a *v* = 35 sin 314 *t* volt supply.

A

$R = 560\,\Omega; V_m = 35\text{V}$

The r.m.s. value for the voltage, $V = 0.707\,V_m$ volt

so, $V = 0.707 \times 35 = 24.75$ V

$$P = \frac{V^2}{R} \text{ watt} = \frac{24.75^2}{560}$$

therefore, $P = 1.09$ W **Ans**

1.2　Pure Inductance

A pure inductor is one which possesses only inductance. It therefore has no electrical resistance or capacitance. Such a device is not practically possible. Since the inductor consists of a coil of wire, then it must possess a finite value of resistance in addition to a very small amount of capacitance. However, let us assume for the moment that an inductor having zero resistance is possible. Consider such a perfect inductor, connected to an a.c. supply, as shown in Fig. 1.5.

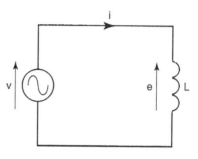

Fig. 1.5

An alternating current will now flow through the circuit. Since the current is continuously changing, then a back emf, e will be induced across the inductor. In this case, e will be exactly equal and opposite to the applied voltage, v. The equation for this back emf is

$$e = -L\frac{di}{dt} \text{ volt}$$

e will have its maximum values when the rate of change of current, di/dt, is at its maximum values. These maximum rates of change occur as the current waveform passes through the zero position. Similarly, e will be zero when the rate of change is zero. This occurs when the current waveform is at its positive and negative peaks.

Thus e will reach its maximum *negative* value when the current waveform has its maximum *positive* slope. Similarly, e will be at its maximum positive value when i reaches its maximum negative slope. Also, since $v = -e$, then the applied voltage waveform will be the 'mirror image' of the waveform for the back emf. These waveforms are shown in Fig. 1.6. From this waveform diagram, it may be seen that the applied voltage, *V leads* the circuit current, *I*, by $\pi/2$ rad, or 90°. The corresponding phasor diagram is shown in Fig. 1.7.

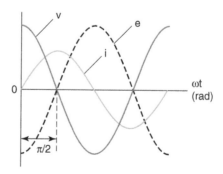

Fig. 1.6

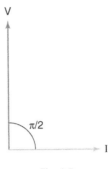

Fig. 1.7

Once more, the instantaneous power is given by the product of instantaneous values of voltage and current. In the first quarter cycle, both v and i are positive quantities. The power is therefore in the positive half of the diagram. In the next quarter cycle, i is still positive, but v is negative. The power waveform is therefore negative. This sequence is repeated every half cycle of the waveform. The average power is therefore zero. Thus, *a perfect inductor dissipates zero power.* These waveforms are shown in Fig. 1.8.

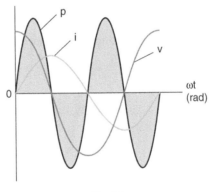

Fig. 1.8

In practical terms, the sequence is as follows. In the first quarter cycle, the magnetic field produced by the coil current, stores energy. In the next quarter cycle, the collapsing field returns all this energy back to the circuit. This sequence is repeated every half cycle. The net result is that the inductor returns as much energy as it receives. Thus no net energy is dissipated, so the power consumption is zero.

1.3   Inductive Reactance

A perfect inductor has no electrical resistance. However, there is some opposition to the flow of a.c. current through it. This opposition is of course due to the back emf induced in the coil. It would be most inconvenient to have to always express this opposition in terms of this emf. It is much better to be able to express the opposition in a quantity that is measured in ohms. This quantity is called the inductive reactance.

Inductive reactance is defined as the opposition offered to the flow of a.c., by a perfect inductor. It is measured in ohms, and the quantity symbol is X_L. The reactance value for an inductor, at a particular frequency, can be determined from a simple equation. This equation may be derived either mathematically or graphically. Both methods will be shown here.

$$e = -L\frac{di}{dt} \text{ volt, and } e = -v$$

$$\text{therefore, } v = L\frac{di}{dt} \text{ volt}$$

Now, $i = I_m \sin \omega t$ amp, so, $v = L\dfrac{d}{dt}(I_m \sin \omega t)$

therefore, $v = \omega L I_m \cos \omega t$

at time $t = 0$, $v = V_m$;   and $\cos \omega t = 1$

hence, $V_m = \omega L I_m$;   and dividing by I_m

$$\frac{V_m}{I_m} = \frac{V}{I} = \omega L \text{ ohm}$$

so, inductive reactance is:

$$X_L = \omega L = 2\pi f L \text{ ohm} \tag{1.3}$$

Alternatively, consider the first quarter cycle of the waveform diagram of Fig. 1.9.

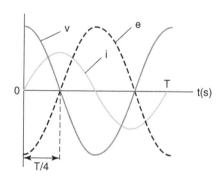

Fig. 1.9

The average emf induced in the coil,

$$e_{av} = \frac{-L\,di}{dt} \text{ volt}$$

$$\text{therefore, } e_{av} = \frac{-L(I_m - 0)}{T/4}; \quad \text{but } T = \frac{1}{f}$$

$$\text{so, } e_{av} = -4LI_m(f) \text{ volt;} \qquad \text{therefore } v_{av} = 4LI_m(f) \text{ volt}$$

$$\text{also, } v_{av} = 0.637V_m = \frac{2V_m}{\pi} \text{ so, } V_m = v_{av} \times \frac{\pi}{2}$$

$$\text{therefore, } V_m = 4LI_m(f) \times \frac{\pi}{2} = 2\pi fLI_m$$

$$\text{hence, } \frac{V_m}{I_m} = \frac{V}{I} = 2\pi fL \text{ ohm}$$

i.e. inductive reactance, $X_L = 2\pi fL = \omega L$ ohm

From equation (1.3), it can be seen that the inductive reactance is directly proportional to both the inductance value and the frequency of the supply. This is logical, since the greater the frequency, the greater the rate of change of current, and the greater the back emf. Figure 1.10 shows the relationship between X_L and frequency f. Notice that the graph goes through the origin $(0, 0)$, confirming that a *perfect* inductor has zero resistance. That is, a frequency of 0 Hz is *d.c.*, so no opposition would be offered to the flow of d.c.

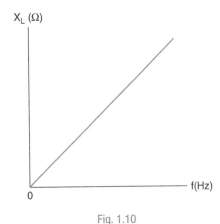

Fig. 1.10

Worked Example 1.2

Q   A pure 20 mH inductor is connected to a 30 V, 50 Hz supply. Calculate (a) the reactance at this frequency, and (b) the resulting current flow.

A

$L = 20 \times 10^{-3}$ H; $V = 30$ V; $f = 50$ Hz

(a) $X_L = 2\pi fL$ ohm $= 2 \times \pi \times 50 \times 20 \times 10^{-3}$

so $X_L = 6.283\,\Omega$ **Ans**

(b) $I = \dfrac{V}{X_L}$ amp $= \dfrac{30}{6.283}$

so $I = 4.77$ A **Ans**

Worked Example 1.3

Q   A current of 250 mA flows through a perfect inductor, when it is connected to a 5 V, 1 kHz supply. Determine the inductance value.

A

$I = 0.25$ A; $V = 5$ V; $f = 1000$ Hz

Firstly, the inductive reactance must be calculated:

$$X_L = \frac{V}{I} \text{ ohm} = \frac{5}{0.25}$$

therefore, $X_L = 20\,\Omega$

Since $X_L = 2\pi f L$ ohm, then $L = \dfrac{X_L}{2\pi f}$ henry

$$L = \frac{20}{2 \times \pi \times 1000}$$

therefore, $L = 3.18$ mH **Ans**

Worked Example 1.4

Q   A coil of inductance 400 μH, and of negligible resistance, is connected to a 5 kHz supply. If the current flow is 15 mA, determine the supply voltage.

A

$L = 400 \times 10^{-6}$ H; $f = 5 \times 10^3$ Hz; $I = 15 \times 10^{-3}$ A

$$X_L = 2\pi f L \text{ ohm} = 2 \times \pi \times 5 \times 10^3 \times 400 \times 10^{-6}$$

so, $X_L = 12.57\,\Omega$

$$V = IX_L \text{ volt} = 15 \times 10^{-3} \times 12.57$$

so, $V = 188.5$ mV **Ans**

1.4   Pure Capacitance

Consider a perfect capacitor, connected to an a.c. s10 upply, as shown in Fig. 1.11. The charge on the capacitor is directly proportional to the p.d. across it. Thus, when the voltage is at its maximum, so too will be the charge, and so on. The waveform for the capacitor charge will therefore be in phase with the voltage. Current is the rate of change of charge. This means that when the rate of change of charge is a maximum, then the current will be at a maximum, and so on. Since the rate of change of charge is maximum as it passes through the zero axis, the current will be at its maximum values at these points. The resulting waveforms are shown in Fig. 1.12. It may therefore be seen that the current now leads the voltage by $\pi/2$ rad, or

90°. To maintain consistency with the inductor previously considered, the circuit current will again be considered as the reference phasor. Thus, we can say that the voltage *lags* the current by π/2 rad. This is illustrated in Fig. 1.13.

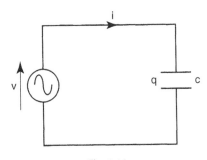

Fig. 1.11

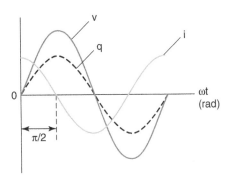

Fig. 1.12

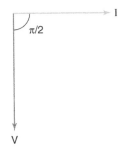

Fig. 1.13

The power at any instant is once more the product of the voltage and current at that instant. When the power waveform is plotted, it will be as shown in Fig. 1.14. Again, it can be seen that the average power is zero. We can therefore conclude that *a perfect capacitor dissipates zero power.* In a similar manner to the inductor, the energy stored in the electric field of the capacitor, in one quarter cycle, is returned to the supply in the next quarter cycle.

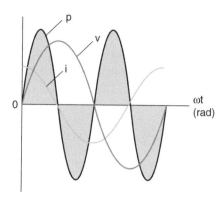

Fig. 1.14

1.5 Capacitive Reactance

The opposition offered to the a.c. current is due to the p.d. developed between the capacitor plates. Again, it is more convenient to refer to this opposition in terms of a quantity measured in ohms. This is known as the capacitative reactance, X_C, which is defined as the opposition offered to the flow of a.c. through a perfect capacitor. The equation for this reactance can be derived in one of two ways. Once more, both will be demonstrated here.

$$q = vC \text{ coulomb}; \quad \text{and } i = \frac{\mathrm{d}q}{\mathrm{d}t} \text{ amp}$$

$$\text{therefore, } i = C\frac{\mathrm{d}V}{\mathrm{d}t}$$

and since $v = V_m \sin \omega t$ volt, then

$$i = C\frac{\mathrm{d}}{\mathrm{d}t}(V_m \sin \omega t)$$

$$= \omega C V_m \cos \omega t$$

when time $t = 0$, $I = I_m$; and $\cos \omega t = 1$ (see Fig. 1.12)

$$\text{therefore, } I_m = \omega C V_m$$

$$\text{and } \frac{V_m}{I_m} = \frac{V}{I} = \frac{1}{\omega C} \text{ ohm}$$

capacity reactance,

$$X_c = \frac{1}{\omega C} = \frac{1}{2\pi f C} \text{ ohm} \qquad (1.4\text{a})$$

Alternatively, consider the waveforms shown in Fig. 1.15.

In general:

average charge = average current × time

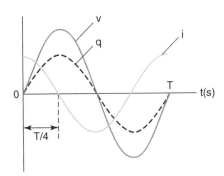

Fig. 1.15

therefore, in first quarter cycle:

$$q = i_{av} \times \frac{T}{4} \text{ coulomb}$$

$$\text{so } q = \frac{i_{av}}{4f}$$

$$\text{but } i_{av} = \frac{2I_m}{\pi} \text{ amp}$$

therefore, $q = I_m 2\pi f$;   and since $q = CV$

$$\text{then, } CV_m = \frac{I_m}{2\pi f}$$

$$\text{and } \frac{V_m}{I_m} = \frac{V}{I} = \frac{1}{2\pi fC} = \frac{1}{\omega C} \text{ ohm}$$

i.e. capacitive reactance,

$$X_C = \frac{1}{2\pi fC} = \frac{1}{\omega C} \text{ ohm} \tag{1.4b}$$

Examination of equation (1.4) shows that the capacitive reactance is *inversely* proportional to both the capacitance and the frequency. Consider the two extremes of frequency. If $f = 0$ Hz (d.c.), then $X_c = $ infinity. This must be correct, since a capacitor does not allow d.c. current to flow through it. At the other extreme, if $f = $ infinity, then $X_c = 0$. The graph of capacitive reactance versus frequency is therefore in the shape of a rectangular hyperbole, as shown in Fig. 1.16.

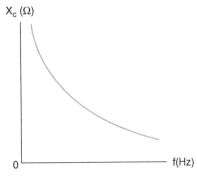

Fig. 1.16

Worked Example 1.5

Q   A 0.47 µF capacitor is connected to a variable frequency signal generator, which provides an output voltage of 25 V. Calculate the current flowing when the frequency is set to (a) 200 Hz, and (b) 4 kHz.

A

$C = 0.47 \times 10^{-6}$ F; $V = 25$ V; $f_1 = 200$ Hz; $f_2 = 4000$ Hz

(a) $X_{C_1} = \dfrac{1}{2\pi f_1 C}$ ohm $= \dfrac{1}{2 \times \pi \times 200 \times 0.47 \times 10^{-6}}$

$X_{C_1} = 1.693$ kΩ

$I_1 = \dfrac{V}{X_C}$ amp $= \dfrac{25}{1693}$

therefore, $I_1 = 14.77$ mA **Ans**

(b) $X_{C_2} = \dfrac{1}{2\pi f_2 C}$ ohm $= \dfrac{1}{2 \times \pi \times 4000 \times 0.47^{-6}}$

$X_{C_2} = 84.66$ Ω

$I_2 = \dfrac{V}{X_{C_2}}$ amp $= \dfrac{25}{84.66}$

therefore, $I_2 = 295.3$ mA **Ans**

Alternatively, we can say that since

$f_2 = 5 \times f_1$, then $X_{C_2} = X_{C_1}/5 = 84.66$ Ω

Hence, $I_2 = 5 \times I_1 = 295.3$ mA **Ans**

Worked Example 1.6

Q   At what frequency will the reactance of a 22 pF capacitor be 500 Ω?

A

$C = 22 \times 10^{-12}$ F; $X_C = 500$ Ω

$$X_C = \dfrac{1}{2\pi f C} \text{ ohm; so } f = \dfrac{1}{2\pi C X_C} \text{ Hz}$$

$$\text{therefore, } f = \dfrac{1}{2 \times \pi \times 500 \times 22 \times 10^{-12}}$$

hence, $f = 14.47$ MHz **Ans**

Summary

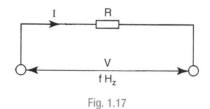

Fig. 1.17

$$I = \dfrac{V}{R} \text{ amp; } V \text{ in phase with } I; \phi = 0; P = I^2 R \text{ watt}$$

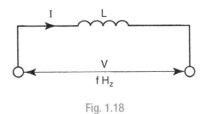

Fig. 1.18

$$I = \frac{V}{X_L} \text{ amp; } V \text{ leads } I \text{ by } 90°; \phi = +\pi/2 \text{ rad } (+90°)$$

$X_L = 2\pi fL$ ohm; P = ZERO watt

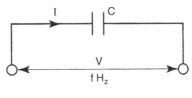

Fig. 1.19

$$I = \frac{V}{X_C} \text{ amp; } V \text{ lags } I \text{ by } 90°; \phi = -\pi/2 \text{ rad } (-90°)$$

$X_C = \dfrac{1}{2\pi fc}$ ohm; P = ZERO watt

CIVIL

In a capacitive (C) circuit	I before V (*I* leads *V*)	V before I in inductive (*V* leads *I*)(L) circuit

The 'keyword' **CIVIL** is a convenient means of remembering whether voltage either leads or lags the circuit current, depending upon whether the circuit is capacitive or inductive. It is particularly useful when making the transition from series to parallel a.c. circuits. In a series circuit, the current is used as the reference phasor. In parallel circuits, the voltage is used as the reference phasor.

1.6   Impedance

This is the total opposition, offered to the flow of a.c. current, by a circuit that contains both resistance and reactance. It is measured in ohms, and has the quantity symbol Z.

$$\text{Thus, } Z = \frac{V}{I} \text{ ohm} \tag{1.5}$$

where V is the circuit applied voltage, and I is the resulting circuit current.

In order to describe the concept of impedance, let us consider a circuit comprising a pure resistor connected in series with a pure inductor. Please note that this arrangement could equally well represent a practical inductor. That is, one which possesses both resistance and inductance. In this latter case, it makes the circuit analysis simpler if we consider the component to consist of the combination of these two separate, pure elements.

1.7   Inductance and Resistance in Series

A pure resistor and a pure inductor are shown connected in series in Fig. 1.20. The circuit current, I, will produce the p.d. V_R across the resistor, due to its resistance, R. Similarly, the p.d. V_L results from the inductor's opposition, the inductive reactance, X_L. Thus, the only circuit quantity that is common, to both the resistor and the inductor, is the circuit current, I. For this reason, the current is chosen as the reference phasor.

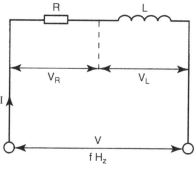

Fig. 1.20

The p.d. across the resistor will be in phase with the current through it ($\phi = 0$). The p.d. across the inductor will lead the current by 90° ($\phi = +90°$). The total applied voltage, V, will be the *phasor sum* of V_R and V_L. This last statement may be considered as the 'a.c. version' of Kirchhoff's voltage law. In other words, the term 'phasor sum' has replaced the term 'algebraic sum', as used in d.c. circuits. The resulting phasor diagram is shown in Fig. 1.21. The angle ϕ, shown

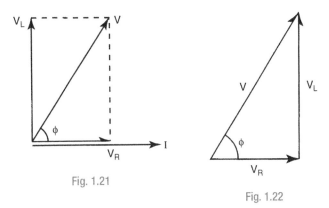

Fig. 1.21

Fig. 1.22

on this diagram, is the angle between the circuit applied voltage, V, and the circuit current, I. It is therefore known as the circuit phase angle.

Phasors are vector quantities. They may therefore be drawn anywhere on the page, provided that they are of appropriate length, and are drawn in the appropriate direction. Thus, the voltage triangle, involving V, V_R and V_L, as in Fig. 1.22, may be 'extracted' from the phasor diagram.

Considering Fig. 1.22, Pythagoras' Theorem may be applied, as follows:

$$V^2 = V_R^2 + V_L^2 \dots\dots\dots\dots[1]$$

Now, $V_R = IR$; $V_L = IX_L$; and $V = IZ$ volt

and substituting these into equation [1] we have:

$$(IZ)^2 = (IR)^2 + (IX_L)^2$$

and dividing through by I^2 we have:

$$Z^2 = R^2 + X_L^2$$

therefore, $Z = \sqrt{R^2 + X_L^2}$ ohm   (1.6)

From the last equation, it may be seen that Z, R and X_L also form a right-angled triangle. This is known as the impedance triangle, and is shown in Fig. 1.23.

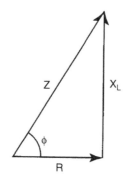

Fig. 1.23

From both the voltage and impedance triangles, the following expressions for the circuit phase angle, ϕ, are obtained:

$$\cos \phi = \frac{R}{Z} = \frac{V_R}{V}$$

$$\text{or, } \sin \phi = \frac{X_L}{Z} = \frac{V_L}{V} \tag{1.7}$$

$$\text{or, } \tan \phi = \frac{X_L}{R} = \frac{V_L}{V_R}$$

Any one of the above three expressions may be used to obtain the value of ϕ. However, as will be seen later, equation (1.7) is most frequently used.

Worked Example 1.7

Q   A resistor of 250 Ω, is connected in series with a 1.5 H inductor, across a 100 V, 50 Hz supply. Calculate (a) the current flowing, (b) the circuit phase angle, (c) the p.d. developed across each component and (d) the power dissipated.

A

$R = 250\,\Omega; L = 1.5\,\text{H}; V = 100\,\text{V}; f = 50\,\text{Hz}$

The relevant circuit and phasor diagrams are shown in Figs. 1.24 and 1.25.

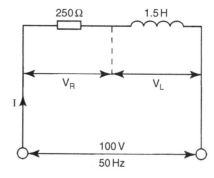

Fig. 1.24

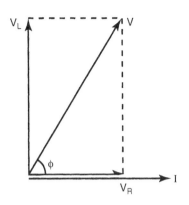

Fig. 1.25

Note: A sketch of these diagrams should normally accompany the written answer.

(a) In order to calculate the current, the impedance must be found. In order to calculate the impedance, the inductive reactance must be known. This then is the starting point in the solution.

$$X_L = 2\pi fL \text{ ohm} = 2 \times \pi \times 50 \times 1.5$$

$$\text{so, } X_L = 471.2\,\Omega$$

$$Z = \sqrt{R^2 + X_L^2} \text{ ohm} = \sqrt{250^2 + 471.2}$$

$$\text{so, } Z = 533.45\,\Omega$$

$$I = \frac{V}{Z}\text{amp} = \frac{100}{533.45}$$

therefore, $I = 187.5$ mA **Ans**

(b)
$$\phi = \cos^{-1}\frac{R}{Z} = \frac{250}{533.45}$$

therefore, $\phi = +62.05°$ or $+1.083$ rad **Ans**

(c)
$$V_R = IR \text{ volt} = 0.1875 \times 250$$

therefore, $V_R = 46.88$ V **Ans**

$$V_L = IX_L \text{ volt} = 0.1875 \times 471.2$$

therefore, $V_L = 88.35$ V **Ans**

(d)
$$P = I^2R \text{ watt} = 0.1875^2 \times 250$$

therefore $P = 8.789$ W **Ans**

Alternatively, the power could be calculated using the V^2/R form, BUT in this case, the voltage must be the actual p.d. across the *resistor*, i.e. V_R. To avoid the possibility of, mistakenly, using V, or V_L, it is better to use the form of equation used above; so use $P = I^2R$ watt.

Worked Example 1.8

Q   A coil, of resistance $35\,\Omega$ and inductance $0.02\,H$, carries a current of $326.5\,mA$, when connected to a $400\,Hz$ a.c. supply. Determine the supply voltage.

A

$$R = 35\,\Omega; L = 0.02\,H; I = 0.3265\,A; f = 400\,Hz$$

$$X_L = 2\pi fL \text{ ohm} = 2 \times \pi \times 400 \times 0.02$$

therefore, $X_L = 50.27\,\Omega$

$$Z = \sqrt{R^2 + X_L^2} \text{ ohm} = \sqrt{35^2 + 50.27^2}$$

therefore, $Z = 61.25\,\Omega$

$$V = IZ \text{ volt} = 0.3265 \times 61.25$$

hence, $V = 20\,V$ **Ans**

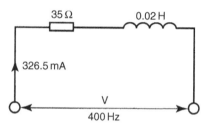

Fig. 1.26

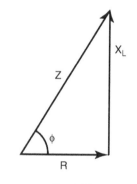

Fig. 1.27

Worked Example 1.9

Q   Two separate tests were carried out on a coil. In the first test, when connected to a $24\,V$ d.c. supply, a current of $2\,A$ flowed through it. In the second test, it was connected to a $24\,V$, $200\,Hz$ supply. In the latter test, the current flow was $1.074\,A$. Calculate the resistance and inductance of the coil.

A

When the coil is connected to a d.c. supply, the only opposition to current flow is the coil resistance. Thus the result of this test can be used to determine the resistance.

$$R = \frac{V}{I} \text{ ohm} = \frac{24}{2}$$

therefore, $R = 12\,\Omega$ **Ans**

When connected to the a.c. supply, the opposition to current flow is the *impedance* of the coil. Thus:

$$Z = \frac{V}{I} \text{ ohm} = \frac{24}{1.074}$$

therefore, $Z = 22.35 \ \Omega$

Since $Z^2 = R^2 + X_L^2$, then $X_L = \sqrt{Z^2 - R^2}$ ohm

therefore, $X_L = \sqrt{22.35^2 - 12^2}$ ohm $= 18.85 \ \Omega$

$$L = \frac{X_L}{2\pi f} \text{ henry} = \frac{18.85}{2 \times \pi \times 200}$$

hence, $L = 15$ mH **Ans**

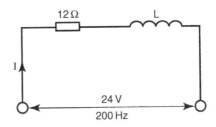

Fig. 1.28

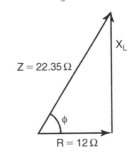

Fig. 1.29

1.8 Resistance and Capacitance in Series

Figure 1.30 shows a pure capacitor and resistor connected in series, across an a.c. supply. Again, being a series circuit, the circuit current is common to both components. Each will have a p.d. developed. In this case however, the p.d. across the capacitor will *lag* the current by 90°.

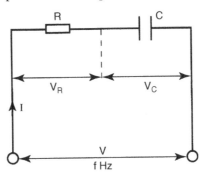

Fig. 1.30

The resulting phasor diagram is shown in Fig. 1.31.

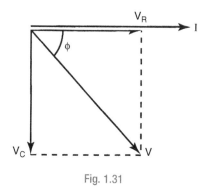

Fig. 1.31

Exactly the same techniques used previously, for the inductor–resistor circuit, may be used here. Thus, both voltage and impedance triangles can be derived. These are shown in Figs. 1.32 and 1.33.

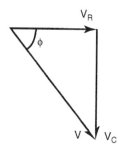

Fig. 1.32

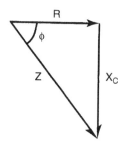

Fig. 1.33

The following equations result:

$$Z = \sqrt{R^2 + X_C^2} \text{ ohm} \qquad (1.8)$$

$$\phi = \cos^{-1}\frac{R}{Z}$$

$$\sin \phi = \frac{X_C}{Z} = \frac{V_C}{V} \qquad (1.9)$$

$$\text{and } \tan \phi = \frac{X_C}{R} = \frac{V_C}{V_R}$$

Worked Example 1.10

Q   A 22 nF capacitor, and a 3.9 kΩ resistor, are connected in series across a 40 V, 1 kHz supply. Determine, (a) the circuit current, (b) the circuit phase angle and (c) the power dissipated.

A

$C = 22 \times 10^{-9}\,\text{F}; R = 3900\,\Omega; V = 40\,\text{V}; f = 10^3\,\text{Hz}$

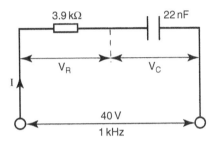

Fig. 1.34

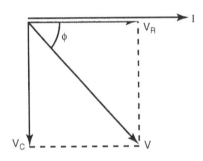

Fig. 1.35

(a) $X_c = \dfrac{1}{2\pi f C}\,\text{ohm} = \dfrac{1}{2 \times \pi \times 10^3 \times 22 \times 10^{-9}}$

so, $X_C = 7.234\,\text{k}\Omega$

$Z\sqrt{R^2 + X_C^2}\,\text{ohm} = \sqrt{(3.9 \times 10^3)^2 + (7.234 \times 10^3)^2}$

hence, $Z = 8.218\,\text{k}\Omega$

$I = \dfrac{V}{Z}\,\text{amp} = \dfrac{40}{8.218 \times 10^3}$

therefore, $I = 4.87\,\text{mA}$ **Ans**

(b) $\phi = \cos^{-1}\dfrac{R}{Z} = \dfrac{3.9 \times 10^3}{8.218 \times 10^3}$

therefore, $\phi = -61.67°$ or $-1.076\,\text{rad}$ **Ans**

(c) $P = I^2 R\,\text{watt} = (4.87 \times 10^{-3})^2 \times 3.9 \times 10^3$

therefore, $P = 92.5\,\text{mW}$ **Ans**

Worked Example 1.11

Q   A 470 Ω resistor is connected in series with a 0.02 μF capacitor, across a 15 V a.c. supply. If the current is 28 mA, calculate (a) the frequency of the supply, and (b) the p.d. across the capacitor.

A

$$C = 0.02 \times 10^{-6}\,\text{F}; R = 470\,\Omega; V = 15\text{V}; I = 0.028\,\text{A}$$

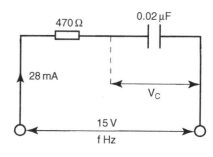

Fig. 1.36

(a)
$$Z = \frac{V}{I}\ \text{ohm} = \frac{15}{0.028}$$

therefore, $Z = 535.7\ \Omega$

Now, $X_C = \sqrt{Z^2 - R^2}\ \text{ohm} = \sqrt{535.7^2 - 470^2}$

hence, $X_C = 257.08\ \Omega$

Since $X_C = \dfrac{1}{2\pi fC}\ \text{ohm}$, then $f = \dfrac{1}{2\pi C X_c}\ \text{Hz}$

therefore, $f = \dfrac{1}{2 \times \pi \times 257.08 \times 0.02 \times 10^{-6}}$

hence, $f = 30.95\ \text{kHz}$ **Ans**

(b)
$$V_C = I X_C\ \text{volt} = 0.028 \times 257.08$$

therefore, $V_C = 7.2\ \text{V}$ **Ans**

1.9   Resistance, Inductance and Capacitance in Series

These three elements, connected in series, are shown in Fig. 1.37. Of the three p.d.s, V_R will be in phase with the current I, V_L will lead I by

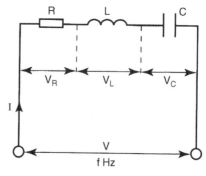

Fig. 1.37

$90°$, and V_C will lag I by $90°$. The associated phasor diagram is shown in Fig. 1.38.

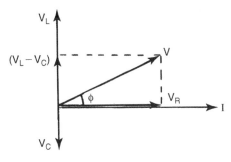

Fig. 1.38

The applied voltage V is the phasor sum of the circuit p.d.s. These p.d.s form horizontal and vertical components. It may be seen that the only horizontal voltage component is V_R. The vertical components, V_L and V_C, are in direct opposition to each other. The resulting vertical component is therefore $(V_L - V_C)$. Applying Pythagoras' theorem, we can say that:

$$V^2 = V_R^2 + (V_L - V_C)^2$$

but, $V = IZ$, $V_R = IR$, $V_L = IX_L$ and $V_C = IX_C$ volt

therefore, $(IZ)^2 = (IR)^2 + (IX_L - IX_C)^2$

hence, $Z^2 = R^2 + (X_L - X_C)^2$

$$\text{and, } Z = \sqrt{R^2 + (X_L - X_C)^2} \text{ ohm} \qquad (1.10)$$

The associated impedance triangle is shown in Fig. 1.39. Note that if $X_C > X_L$, then the circuit phase angle ϕ will be lagging, instead of leading as shown.

$$\cos \phi = \frac{R}{Z}; \quad \sin \phi = \frac{X_L - X_C}{Z};$$

$$\text{and } \tan \phi = \frac{X_L - X_C}{R}$$

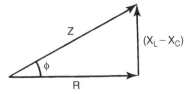

Fig. 1.39

Worked Example 1.12

Q   A coil of resistance $8\,\Omega$ and inductance $150\,\text{mH}$, is connected in series with a $100\,\mu\text{F}$ capacitor, across a $240\,\text{V}$, $50\,\text{Hz}$ a.c. supply. Calculate (a) the circuit current, (b) the circuit phase angle, (c) the p.d. across the coil, (d) the p.d. across the capacitor, and (e) the power dissipated.

A

$R = 8\,\Omega; L = 0.15\,\text{H}; C = 10^{-4}\,\text{F}; V = 240\,\text{V}; f = 50\,\text{Hz}$

The circuit diagram is shown in Fig. 1.40. Note that the coil has been considered as the combination of a pure resistor and a pure inductor.

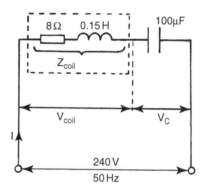

Fig. 1.40

(a) $X_L = 2\pi fL\ \text{ohm} = 2 \times \pi \times 50 \times 0.15$

so, $X_L = 47.12\,\Omega$

$$X_C = \frac{1}{2\pi fC}\ \text{ohm} = \frac{1}{2 \times \pi \times 50 \times 10^{-4}}$$

so $X_C = 31.83\,\Omega$

$$Z = \sqrt{R^2 + (X_L - X_C)^2}\ \text{ohm}$$

$$= \sqrt{18^2 + (47.12 - 31.83)^2}$$

therefore $Z = 17.26\,\Omega$

$$I = \frac{V}{Z}\ \text{amp} = \frac{240}{17.26}$$

therefore $I = 13.91\,\text{A}$ **Ans**

(b) $$\phi = \cos^{-1}\frac{R}{Z} = \cos^{-1}\frac{8}{17.26}$$

therefore, $\phi = +62.38°$ or $+1.089\,\text{rad}$ **Ans**

(c)   The coil has both inductance and reactance. The coil itself therefore possesses impedance, Z_{coil} ohm. The p.d. across the coil is therefore due to its impedance, and NOT only R or X_L.

$$Z_{coil} = \sqrt{R^2 + X_L^2}\ \text{ohm} = \sqrt{8^2 + 47.12^2}$$

hence, $Z_{coil} = 47.8\,\Omega$

$$V_{coil} = IZ_{coil}\ \text{volt} = 13.91 \times 47.8$$

therefore, $V_{coil} = 664.9\,\text{V}$ **Ans**

(d) $V_C = IX_C$ volt $= 13.91 \times 31.83$

hence, $V_C = 442.8$ V **Ans**

(e) $P = I^2R$ watt $= 13.91^2 \times 8$

therefore, $P = 1.547$ kW **Ans**

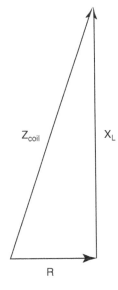

Fig. 1.41

Note: From the above calculations, we have the result that the p.d.s, across both the coil and the capacitor, are greater than the applied voltage, *V*. At first, this may seem to be incorrect, since it would not be possible in a d.c. circuit. In a d.c. circuit, the applied voltage equals the *algebraic* sum of the p.d.s. In the a.c. circuit, the applied voltage is the *phasor* sum of the p.d.s. Thus, it is perfectly possible for the p.d. across the capacitor, or the coil, or both, to be greater than the applied voltage. This is because the total vertical component of voltage is the *difference* between V_L and V_C. This condition is therefore possible only when *all three* components are in series.

1.10 Power in the A.C. Circuit

It has been shown that power is dissipated only by the *resistance* in the circuit. Pure inductance and capacitance do not dissipate any net power. Considering a resistor, we know that the p.d. across it is in phase with the current flowing through it ($\phi = 0°$). For both the inductor and the capacitor, $\phi = 90°$. Thus, we can conclude that only the components of voltage and current, that are in phase with each other, dissipate power. Consider the phasor diagram for an *R-L* series circuit, as shown in Fig. 1.42.

From this diagram, it can be seen that the only two components of current and voltage that are in phase with each other are *I* and V_R.

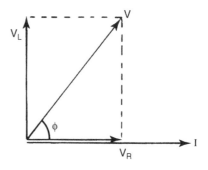

Fig. 1.42

Thus, $P = V_R I$ watt

but, $V_R = V \cos \phi$ volt

therefore, $P = VI \cos \phi$ watt   (1.11)

1.11   Power Factor

This is defined as the ratio of the true power, to the product of the r.m.s. values of applied voltage and current (apparent power).

Thus, power factor (p.f.) $= \dfrac{VI \cos \phi}{VI}$

hence, p.f. $= \cos \phi$ (1.12)

To illustrate the concepts of true power and apparent power, consider the $R\text{-}L$ circuit, shown in Fig. 1.43, where the components are inside a box, and therefore not visible. Let the applied voltage V and the current I be measured, using a voltmeter and ammeter respectively. Taking these two meter readings, it would *appear* that the circuit within the box consumes a power of VI watt.

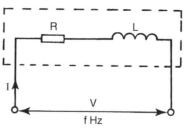

Fig. 1.43

However, if a wattmeter were connected, this would indicate the true power. This wattmeter reading would be found to be less than the product of the ammeter and voltmeter readings. Since the true and apparent powers are related by the power factor, cos ϕ, then they must form two sides of a right-angled triangle.

1.12   Power Triangle

The true power is measured in watts (W). The apparent power is measured in volt-ampere (VA). The third side of the triangle is known as the reactive component of power, and is measured in reactive volt-ampere (VAr). The quantity symbols are P, S and Q respectively. These are shown in Fig. 1.44.

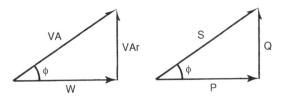

Fig. 1.44

Worked Example 1.13

Q   An R-C circuit consisting of a 4.7 μF capacitor in series with a 200 Ω resistor, is connected to a 250 V, 50 Hz supply. Determine, (a) the current, (b) the power factor, and (c) the values for true, apparent and reactive powers.

A

$C = 4.7 \times 10^{-6}$ F; $R = 200\,\Omega$; $V = 250$ V; $f = 50$ Hz

(a)
$$X_C = \frac{1}{2\pi f C}\ \text{ohm} = \frac{1}{2 \times \pi \times 50 \times 4.7 \times 10^{-6}}$$

so, $X_C = 677.26\,\Omega$

$$Z = \sqrt{R^2 - X_C^2}\ \text{ohm} = \sqrt{200^2 + 677.26^2}$$

therefore, $Z = 706.2\,\Omega$

$$I = \frac{V}{Z}\ \text{amp} = \frac{250}{706.2}$$

hence $I = 0.354$ A **Ans**

(b)
$$\text{p.f.} = \cos\phi = \frac{R}{Z} = \frac{200}{706.2}$$

therefore, p.f. = 0.283 lagging **Ans**

(c) $P = VI\cos\phi\ \text{watt} = 250 \times 0.354 \times 0.283$

therefore, $P = 25.06$ W **Ans**

From the power triangle (Fig. 1.45)

$$S = VI\ \text{volt-ampere} = 250 \times 0.354$$

therefore, $S = 88.5$ VA **Ans**

$$Q = S\sin\phi\ \text{reactive volt-ampere}$$

$$= 88.5 \times 0.9591$$

hence $Q = 84.88$ VAr **Ans**

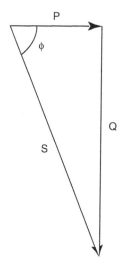

Fig. 1.45

Worked Example 1.14

Q   The stator winding of a single-phase a.c. motor consumes 250 W, at a leading power factor of 0.537, when connected to a 240 V, 50 Hz supply. Calculate (a) the current drawn, and (b) the resistance and inductance of the winding.

A

$P = 250\,\text{W}; \cos \phi = +0.537; V = 240\,\text{V}; f = 50\,\text{Hz}$

(a) $P = VI \cos \phi$ watt

$$\text{so, } I = \frac{P}{V \cos \phi}\,\text{amp} = \frac{250}{240 \times 0.537}$$

therefore, $I = 1.94$ A **Ans**

(b) $Z = \dfrac{V}{I}\,\text{ohm} = \dfrac{240}{1.94}$

so $Z = 123.72\ \Omega$

$\phi = \cos^{-1} 0.537 = 57.52°$; so $\sin \phi = 0.8436$

The impedance triangle is shown in Fig. 1.46. From this:

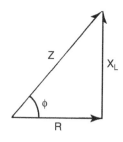

Fig. 1.46

$$R = Z \cos \phi = 123.72 \times 0.537$$

therefore, $R = 66.44 \ \Omega$ **Ans**

$$X_L = Z \sin \phi = 123.72 \times 0.8436$$

so $X_L = 104.37 \ \Omega$

$$L = \frac{X_L}{2\pi f} \ \text{henry} = \frac{104.37}{2 \times \pi \times 50}$$

therefore, $L = 0.332$ H **Ans**

Worked Example 1.15

Q A circuit draws a power of 2.5 kW at a leading power factor of 0.866. Calculate the apparent power and the reactive component of power.

A

$P = 2500\,\text{W}; \cos \phi = 0.866$

Phase angle, $\phi = \cos^{-1} 0.866 = 30°$, and the power triangle will be as shown in Fig. 1.47.

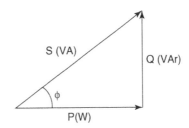

Fig. 1.47

$$\cos \phi = \frac{P}{S}$$

so apparent power, $S = \dfrac{P}{\cos \phi}$ volt-ampere

$$= \frac{2500}{0.866}$$

hence, $S = 2.89$ kVA **Ans**

$$\tan \phi = \frac{Q}{P}$$

so, reactive component, $Q = P \tan \phi$ volt-ampere reactive

$$= 2.5 \times \tan 30° \ \text{kVAr}$$

therefore, $Q = 1.44$ kVAr **Ans**

1.13   Series Resonance

Consider a series *R-L-C* circuit, connected to a constant amplitude, variable frequency supply, as in Fig. 1.48. Let the frequency be initially set at a low value. In this case, the capacitive reactance will be relatively large. The inductive reactance will be relatively small. The resistance will remain constant throughout. The relevant phasor diagram is shown in Fig. 1.49(a).

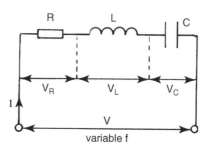

Fig. 1.48

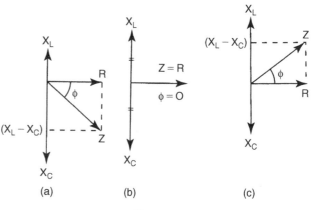

Fig. 1.49

Let the frequency of the supply now be increased. As this is done, so X_C will decrease, and X_L will increase. At one particular frequency, the values of X_C and X_L will be equal. This is known as the resonant frequency, f_0, and the general condition is known as resonance. This effect is illustrated in Fig. 1.49(b).

If the frequency continues to be increased, then X_L will continue to increase, whilst X_C continues to decrease. This is shown in Fig. 1.49(c).

Considering the circuit phase angle, it can be seen that it has, in turn, changed from lagging to zero (at resonance) to leading. This sequence, in terms of the voltage phasor diagram, is shown in Fig. 1.50(a), (b) and (c).

At the resonant frequency, the applied voltage is in phase with the circuit current. Additionally, under this condition, $V = V_R$. Thus, from

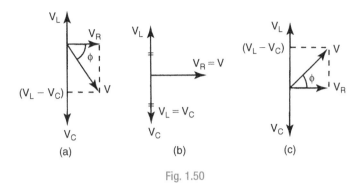

Fig. 1.50

the point of view of the supply, the circuit behaves as if it consists only of the resistor. This is because, at f_0, $X_L = X_C$. These reactances therefore effectively 'cancel' each other. This is illustrated below:

$$Z = \sqrt{R^2 + (X_L - X_C)^2} \text{ ohm}$$
$$\text{and, at } f_0, X_L = X_C$$
$$\text{therefore, } Z = \sqrt{R^2}$$

hence, at resonance $Z = R$ ohm   (1.13)

It should be apparent that this is the minimum possible value for the circuit impedance. So, at resonance, the circuit current must reach its maximum possible value. The variation of the reactances, and hence the impedance, is illustrated in Fig. 1.51. The variation of the circuit current, with frequency, is shown in Fig. 1.52.

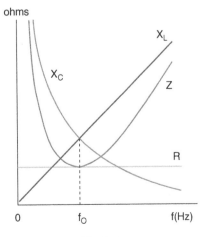

Fig. 1.51

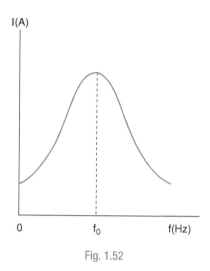

Fig. 1.52

The value for the resonant frequency, for a given circuit, can be determined as follows:

$$\text{At } f_0, \; \omega_0 L = \frac{1}{\omega_0 C} \text{ ohm}$$

$$\text{therefore, } \omega_0^2 LC = 1$$

$$\text{and, } \omega_0^2 = \frac{1}{LC}$$

$$\text{so, } \omega_0 = \frac{1}{\sqrt{LC}} \text{ rad/s}$$

$$\text{hence, } f_0 = \frac{1}{2\pi\sqrt{LC}} \text{ hertz} \qquad (1.14)$$

Note: Under resonant conditions, the p.d.s developed across the capacitor and inductor can be many times greater than the supply voltage. This could have serious consequences for the components used. The dielectric strength of the capacitor may well be exceeded. The insulation on the coil could also prove to be insufficient. For these reasons, the resonant condition is normally avoided, when dealing with 'power' circuits (i.e. 50 Hz supplies). At higher frequencies (e.g. radio frequencies), the applied voltages are normally fairly small. At these frequencies, resonance does not pose the same problem. Indeed, at these frequencies, the resonance effect can be put to good use. This aspect is dealt with in Chapter 2.

Worked Example 1.16

Q   A coil of resistance 10 Ω, and inductance 1.013 H, is connected in series with a 10 μF capacitor. Calculate (a) the resonant frequency, (b) the circuit current, when connected to a 240 V, 50 Hz supply, and (c) the p.d. developed across the capacitor.

A

$$R = 10\,\Omega; L = 1.013\,\text{H}; C = 10 \times 10^{-6}\,\text{F}; V = 240\text{V}$$

(a) $$f_0 = \frac{1}{2\pi\sqrt{LC}} \text{ hertz} = \frac{1}{2\pi\sqrt{1.013 \times 10^{-5}}}$$

thus, $f_0 = 50$ Hz **Ans**

(b) Since the supply frequency is 50 Hz, then the circuit will resonate. This means that the only opposition to current flow is the *resistance* of the coil, R ohm.

Therefore, $I = \dfrac{V}{R}$ amp $= \dfrac{240}{10}$

hence, $I = 24$ A **Ans**

(c) $$X_C = \frac{1}{2\pi f C} \text{ ohm} = \frac{1}{2 \times \pi \times 50 \times 10^{-5}}$$

so, $X_C = 318.31\ \Omega$

$V_C = I X_C$ volt $= 24 \times 318.31$

therefore, $V_C = 7639$ V **Ans**

The above example clearly illustrates the problem of resonance at power frequency. For example, suppose the capacitor dielectric has a thickness of 1 mm. The peak voltage between the plates will be $\sqrt{2} \times 7639$ V $= 10\,803$ V. The *minimum* dielectric strength required would therefore be $10\,803/10^{-3} = 10.8$ MV/m.

If the frequency was increased to 200 Hz, the applied voltage remaining at 240 V, the current would be reduced to 0.518 A, and the p.d. across the capacitor to 46.25 V.

Summary of Equations

Resistance: $I = \dfrac{V}{R}$ amp; $P = I^2R$ watt; $\phi = 0°$

Inductance: Reactance, $X_L = 2\pi f L$ ohm; $I = \dfrac{V}{X_L}$ amp; $P = 0$W; $\phi = +90°$ or $+\dfrac{\pi}{2}$ rad

Capacitance: Reactance, $X_C = \dfrac{1}{2 f \pi C}$ ohm; $I = \dfrac{V}{X_C}$ amp; $P = 0$W; $\phi = -90°$ or $-\dfrac{\pi}{2}$ rad

Impedance: $Z = \sqrt{R^2 + (X_L - X_C)^2}$ ohm; $\cos\phi = \dfrac{R}{Z}$; $P = VI\cos\phi = I^2R$ watt

$I = \dfrac{V}{Z}$ amp

Power triangle: Power factor $= \cos \phi$

True power in W; apparent power in VA; reactive 'power' in VAr.

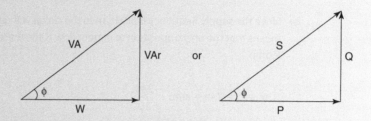

Series resonance: Resonant frequency, $f_0 = \dfrac{1}{2\pi\sqrt{LC}}$ hertz

$X_L = X_C$ ohm; $Z = R$ ohm; $I = \dfrac{V}{R}$ amp (max. possible value)

Assignment Questions

1 Calculate the reactance of a perfect capacitor, of 0.47 µF, at the following frequencies. (a) 50 Hz, (b) 200 Hz, (c) 5 kHz, and (d) 300 kHz.

2 Determine the reactance of a perfect inductor, of 0.5 H, at the following frequencies. (a) 40 Hz, (b) 600 Hz, (c) 4 kHz, and (d) 1 MHz.

3 A pure inductor has a reactance of 125 Ω, when connected to a 5 kHz supply. Calculate the inductance.

4 A perfect capacitor has a reactance of 100 Ω, when connected to a 60 Hz supply. Determine its capacitance.

5 A 250 V, 50 Hz supply is connected to a pure inductor. The resulting current is 1.5 A. Calculate the value of the inductance.

6 The current drawn by a perfect capacitor is 250 mA, when connected to a 25 V, 1.5 kHz supply. Determine the capacitance value.

7 Determine the frequency at which a 40 mH inductor will have a reactance of 60 Ω.

8 At what frequency will a 5000 pF capacitor have a reactance of 2 kΩ?

9 A 24 V, 400 Hz a.c. supply is connected, in turn, to the following components. (a) a 560 Ω resistor, (b) a pure inductor, of 20 mH, and (c) a pure capacitor of 220 pF. For each case, calculate the current flow and power dissipation.

10 Determine the impedance of a coil, which has a resistance of 15 Ω and a reactance of 20 Ω.

11 A coil of inductance 85 mH, and resistance 75 Ω, is connected to a 120 V, 200 Hz supply. Calculate (a) the impedance, (b) the circuit current, (c) the circuit phase angle and power factor, and (d) the power dissipated.

12 For the circuit shown in Fig. 1.53, determine (a) the supply voltage, and (b) the circuit phase angle.

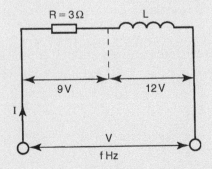

R = 3 Ω　　　　L

9 V　　　　12 V

I

V

f Hz

Fig. 1.53

13 For the circuit of Fig. 1.53, assuming a supply frequency of 1.5 kHz, calculate (a) the current, (b) the inductance and (c) the power dissipated.

14 A coil takes a current of 4.5 A, when connected to a 24 V d.c. supply. When connected to a 24 V, 50 Hz supply, it takes a current of 2.6 A. Explain why the current in the second case is less than in the first. Hence, calculate (a) the resistance of the coil and (b) its inductance.

15 A voltage of 45 V, 80 Hz, is applied across a circuit consisting of a capacitor in series with a resistor. Determine the p.d. across the capacitor, if the p.d. across the resistor is 20 V.

16 A 75 Ω resistor is connected in series with a 15 µF capacitor. If this circuit is supplied at 180 V, 100 Hz, calculate (a) the impedance, (b) the current, (c) the power factor and (d) the power dissipated.

17 For the circuit shown in Fig. 1.54, the reactance of the capacitor is 1600 Ω. Determine (a) the supply voltage, (b) the circuit current, (c) the supply frequency and (d) the value of the resistance.

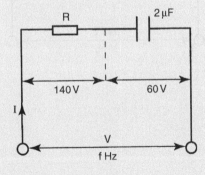

R　　　　　　2 µF

140 V　　　　60 V

I

V

f Hz

Fig. 1.54

18 An alternating voltage of $v = 180 \sin 628.4t$ volt is applied to a circuit containing a 39 Ω resistor, connected in series with a 47 µF capacitor. Calculate (a) the circuit current, (b) the p.d.s across the resistor and capacitor, (c) the power dissipated, and (d) the power factor.

Assignment Questions

19   For the circuit shown in Fig. 1.55, determine (a) the impedance, (b) the current, (c) the phase angle, (d) the p.d. across each component.

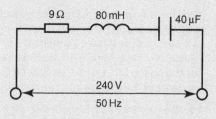

Fig. 1.55

20   The circuit shown in Fig. 1.56 dissipates 162.5 W. Calculate (a) the circuit current, (b) the p.d. across the coil, (c) the p.d. across the capacitor, and (d) the phase angle of the circuit.

21   Determine the resonant frequency for the circuit shown in Fig. 1.56. What would be the current flowing under this condition? Explain why this condition would be avoided at power frequencies.

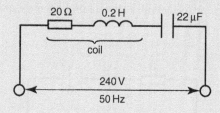

Fig. 1.56

22   An *R-L-C* circuit resonates at a frequency of 10 kHz. If the capacitor has a value of 400 pF, calculate the value of the inductor.

23   A transformer is rated at 15 kVA, at a lagging power factor of 0.75. Determine the rated power output, in kW.

24   A load dissipates 40 kW, at a power factor of 0.6 lagging. Calculate the apparent power, and its reactive component.

25   The power consumed by an *R-C* circuit, when connected to a 100 V, 2 kHz supply, is 800 W. If the circuit current is 12 A, calculate (a) the impedance, (b) the resistance, (c) the reactance, (d) the capacitance, (e) the power factor, and (f) the circuit phase angle.

26   An a.c. motor draws a current of 10 A, when connected to a 240 V, 50 Hz supply. If the power factor is 0.6 lagging, calculate (a) the true power, apparent power, and reactive component of power, (b) the resistance and inductance of the motor winding.

Suggested Practical Assignments

Assignment 1

To determine the variation of capacitive reactance, with variation of frequency.

Apparatus:

1 × 1 μF capacitor
1 × signal generator
1 × ammeter
1 × voltmeter

Method:

1   Connect the capacitor and ammeter in series, across the output terminals of the signal generator.
2   Set the output voltage to 10V (as measured across the capacitor), and the frequency to 300 Hz. Measure the current flow. Maintain this voltage value throughout.
3   Alter the frequency, in 100 Hz steps, up to 1 kHz. Record the ammeter and voltmeter readings at each step.
4   Alter the frequency, in 500 Hz steps, up to 4 kHz. Record the meter readings at each step.
5   From your tabulated values of I and V, calculate and tabulate the values for X_C.
6   Plot two graphs. One graph to be X_C versus frequency; the other to be X_C versus reciprocal of frequency $(1/f)$.
7   Write an assignment report, and from your results, state the relationship between capacitive reactance and frequency.

Assignment 2

To investigate the reactance and impedance of a coil.

Apparatus:

1 × coil
1 × signal generator
1 × d.c. supply
1 × ammeter
1 × voltmeter

Method:

1   Connect the coil to the d.c. supply.
2   Increase the voltage applied to the coil, in steps, and note the current at each step.
3   Tabulate these meter readings, and plot a graph of V versus I. From this graph, determine the resistance of the coil.
4   Connect the coil to the signal generator, set to a frequency of 1 kHz.
5   Vary the output voltage of the signal generator, in steps, and record the voltage and current readings at each step.
6   From your tabulated values, plot a graph of V versus I.
7   From this graph, determine the impedance of the coil. Hence, determine the reactance of the coil at 1 kHz, and its inductance.
8   Complete an assignment report.

Assignment 3

To investigate the phase relationship between current and voltage in a capacitor circuit.

Apparatus:

1 × double-beam oscilloscope
1 × signal generator
1 × 0.22 µF capacitor
1 × 1 kΩ resistor

Method:

1   Connect the resistor and capacitor, in series, across the output terminals of the signal generator.
2   Connect the oscilloscope, such that one channel monitors the p.d. across the capacitor, and the other the p.d. across the resistor.
3   Set the frequency to 1 kHz, and sketch the two waveforms observed.
4   Make a careful note of the time interval between the two waveforms, and hence calculate the phase difference between them.

Note: The current through a resistor is in phase with the p.d. across it. Thus, the waveform for the resistor p.d., may be taken to represent the circuit current.

Assignment 4

To investigate the phase relationship, between current and voltage, for an inductor.

Apparatus:

1 × low resistance coil
1 × 5 kΩ resistor
1 × signal generator
1 × double-beam oscilloscope

Method:

1   Connect the resistor and coil, in series, across the output terminals of the signal generator. Set the frequency to 10 kHz.
2   Using the two channels of the oscilloscope, monitor the p.d.s across the resistor and coil.
3   Carefully note the time interval between the two waveforms.
4   Calculate the phase angle between the two waveforms, and sketch the waveforms.

Note: The waveform of p.d. across the resistor, may again be taken to represent the circuit current. Provided that the coil resistance is very low compared to the reactance, then the coil may be considered to be almost ideal.

Assignment 5

To investigate the effect of series resonance.

Apparatus:

1 × signal generator
1 × inductor (of known value)
1 × capacitor (of known value)
1 × voltmeter
1 × ammeter

Method:

1   Connect the capacitor, inductor and ammeter in series, across the terminals of the signal generator.
2   Calculate the theoretical resonant frequency of the circuit.
3   Set the frequency to $f_0/10$. Record the circuit current, and the capacitor p.d. Ensure that the output voltage of the signal generator is maintained constant throughout.

4   Increase the frequency, in discrete steps, up to $f_0 \times 10$. Record the current and voltage readings at each step.

Note: Take 'extra' readings at, and on either side of the resonant frequency f_0 Hz.

5   Plot a graph of circuit current versus frequency. Compare the actual resonant frequency to the calculated value.
6   Complete an assignment report.

Chapter 2

Single-Phase Parallel A.C. Circuits

Learning Outcomes

This chapter deals with the solution of parallel circuits, and the effects of both series and parallel resonance. Also included are the technique of power factor correction, and the concepts of tuned circuits and filters.

On completion of the chapter, you should be able to:

1 Determine the current flows, p.d.s and power dissipation in parallel a.c. circuit configurations.
2 State the conditions for resonance in both series and parallel circuits, and apply the formulae to obtain the resonant frequency in each case.
3 Define and calculate circuit Q-factor.
4 Appreciate the need for power factor correction, and carry out the relevant calculations.
5 Understand the effect of turned (resonant) circuits on circuit selectivity and bandwidth, and their application to filter circuits.

2.1 Summary of Series A.C. Circuits and Equations

The solution of parallel a.c. circuits is a simple extension of the methods used for series circuits. It is therefore sensible, at this stage, to summarise what has been learned about series circuits.

1 In a series circuit, the current is common to all of the series elements. For this reason the current is chosen as the reference phasor.
2 In a series circuit, the *phasor* sum of the p.d.s equals the total applied voltage.
3 Reactance is the opposition offered, to the flow of a.c., by a pure inductor or capacitor. Reactance is measured in ohms, and the symbols are X_L and X_C.

4   Reactance depends upon the frequency of the supply, such that:

$$X_L = 2\pi f_L \text{ ohm}$$

$$\text{and } X_C = \frac{1}{2\pi f C} \text{ ohm}$$

5   Impedance is the overall opposition offered to the flow of a.c., in a circuit that contains both resistance and reactance. Impedance has the symbol Z, and is measured in ohms.

6   $Z = \dfrac{V}{I} = \sqrt{R^2 + (X_L - X_C)^2} \text{ ohm}$

7   Power is dissipated *only* by a resistive component. The power may be calculated from either:

$$P = I^2 R \text{ or } P = VI \cos \phi \text{ watt}$$

where ϕ is the circuit phase angle, and $\cos \phi$ is the circuit power factor.

8   For a pure resistor,

$$\phi = 0° \quad V_R \text{ in phase with } I$$

For a pure inductor,

$$\phi = +90° \quad V_L \text{ leads } I \text{ by } 90°$$

For a pure capacitor,

$$\phi = -90° \quad V_C \text{ lags } I \text{ by } 90°$$

the above being summed up by the 'keyword' CIVIL. Typical circuit and phasor diagrams are shown in Figs. 2.1 and 2.2.

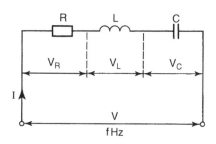

Fig. 2.1

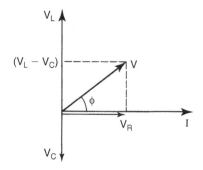

Fig. 2.2

2.2   The *R-C* Parallel Circuit

Consider a pure resistor and a pure capacitor, connected in parallel across an a.c. supply, as shown in Fig. 2.3. Since both components are connected directly to the same supply terminals, then the *supply voltage, V,* is common to them both. On the other hand, each component will draw a current dependent upon its opposition offered.

$$\text{i.e. } I_1 = \frac{V}{R} \text{ amp, and } I_2 = \frac{V}{X_C} \text{ amp}$$

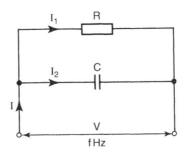

Fig. 2.3

However, both I_1 and I_2 are supplied from the a.c. source. Thus the total current drawn from the supply, *I* is the **PHASOR SUM** of I_1 and I_2. The resulting phasor diagram, as shown in Fig. 2.4, therefore uses the applied voltage *V* as the reference phasor. From this diagram, it may be seen that

$$I = \sqrt{I_1^2 + I_2^2} \text{ amp}$$

$$\text{and } \cos \phi = \frac{I_1}{I} \text{ or } \tan \phi = \frac{I_2}{I_1}$$

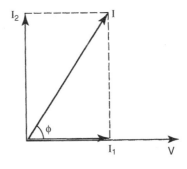

Fig. 2.4

Note: In a parallel circuit, the equation $Z = \sqrt{R^2 + X_C^2}$ **CANNOT BE USED** to obtain the circuit impedance. The impedance can only be obtained from $Z = V/I$ ohm. Also note the fact that since the current through the capacitor (I_2) *leads* the voltage V, then a capacitor in a parallel circuit results in a leading phase angle, ϕ. This may be confirmed by considering the word CIVIL.

Worked Example 2.1

Q   A 2.2 µF capacitor is connected in parallel with a 1.5 kΩ resistor across a 100 V, 50 Hz supply. Sketch the circuit and phasor diagrams, and calculate (a) the current flow through each component, (b) the current drawn from the supply, (c) the circuit phase angle, and (d) the power dissipated.

A

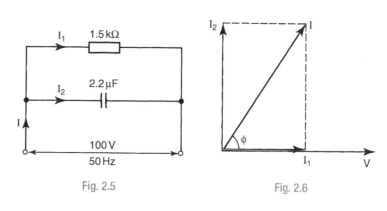

Fig. 2.5   Fig. 2.6

$C = 2.2 \times 10^{-6}$ F; $R = 1500\ \Omega$; $V = 100$ V; $f = 50$ Hz

(a) $I_1 = \dfrac{V}{R}$ amp $= \dfrac{100}{1500}$

so $I_1 = 66.7$ mA **Ans**

$X_C = \dfrac{1}{2\pi f C}$ ohm $= \dfrac{1}{2 \times \pi \times 50 \times 2.2 \times 10^{-6}}$

so $X_C = 1446.9\ \Omega$

$I_2 = \dfrac{V}{X_C}$ amp $= \dfrac{100}{1446.9}$

so $I_2 = 69.1$ mA **Ans**

(b) $I = \sqrt{I_1^2 + I_2^2}$ amp $= \sqrt{66.7^2 + 69.1^2}$ mA

hence $I = 96$ mA **Ans**

(c) $\phi = \cos^{-1}\dfrac{I_1}{I} = \cos^{-1}\dfrac{66.7}{96}$

so $\phi = 46.03°$ leading **Ans**

(d) $P = VI \cos\phi$ watt $= 100 \times 96 \times 10^{-3} \times 0.6943$

so $P = 6.67$ W **Ans**

Alternatively, $P = I_1^2 R$ watt $= (66.7 \times 10^{-3})^2 \times 1500$

so $P = 6.67$ W **Ans**

Worked Example 2.2

Q   For the circuit shown in Fig. 2.7, determine (a) the supply voltage, (b) the supply frequency, and (c) the current drawn from the supply.

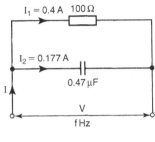

Fig. 2.7

A

(a)   Since the supply is connected directly across the resistor, then the p.d. across $R = V$.

therefore, $V = I_1 R$ volt $= 0.4 \times 100$
so, $V = 40$ V **Ans**

(b) $$X_C = \frac{V}{I_2} \text{ ohm} = \frac{40}{0.177}$$

therefore, $X_C = 225.99\ \Omega$

and, since $X_C = \dfrac{1}{2\pi f C}$ ohm

then, $f = \dfrac{1}{2\pi C X_C}$ hertz

so, $f = \dfrac{1}{2 \times \pi \times 225.99 \times 0.47 \times 10^{-6}}$

hence, $f = 1498$ Hz **Ans**

(c) $I = \sqrt{I_1^2 + I_2^2}$ amp $= \sqrt{0.4^2 + 0.177^2}$

thus, $I = 0.437$ A **Ans**

Worked Example 2.3

Q   A 4.7 μF capacitor is connected in parallel with a 500 Ω resistor across a 50 Hz a.c. supply. If the current through the resistor is 200 mA, calculate (a) the supply voltage, (b) the current through the capacitor, (c) the total current drawn from the supply, and (d) the circuit power factor.

A

$C = 4.7 \times 10^{-6}$ F; $R = 500\ \Omega$; $f = 50$ Hz; $I_1 = 0.2$ A

(a)   Since the applied voltage is common to both the resistor and the capacitor, then

$V = I_1 R$ volt $= 0.2 \times 500$

so, $V = 100$ V **Ans**

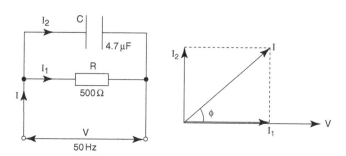

Fig. 2.8

(b) $X_C = \dfrac{1}{2\pi fC}$ ohm $= \dfrac{1}{2\pi \times 50 \times 4.7 \times 10^{-6}}$

so, $X_C = 677.26\ \Omega$ **Ans**

$I_2 = \dfrac{V}{X_C}$ amp $= \dfrac{100}{677.26}$

and $I_2 = 147.7$ mA **Ans**

(c) $I = \sqrt{I_1^2 + I_2^2}$ amp $= \sqrt{0.2^2 + 0.147^2}$

hence, $I = 248.6$ mA **Ans**

(d)   p.f. $= \cos\phi = \dfrac{I_1}{I} = \dfrac{200}{248.6}$

and p.f. $= 0.805$ leading **Ans**

2.3   The *R-L* Parallel Circuit

When a pure inductor is connected in parallel with a resistor, the techniques used to analyse the circuit are the same as for the *C-R* circuit. The only difference will be that the circuit will have a lagging phase angle. The following example illustrates this.

Worked Example 2.4

Q   The circuit of Fig. 2.9 shows a pure inductor connected in parallel with a 200 Ω resistor, across a 50 V, 1 kHz supply. If the current drawn from the supply is 1.03 A at a power factor of 0.2427, determine (a) the current in each branch, (b) the power dissipated, and (c) the value of the inductance.

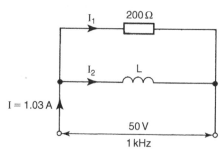

Fig. 2.9

A

$V = 50\,\text{V}; f = 10^3\,\text{Hz}; I = 1.03\,\text{A}; \cos \phi = 0.2427$

The corresponding phasor diagram is shown in Fig. 2.10.

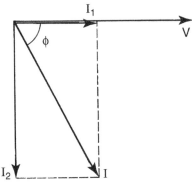

Fig. 2.10

(a) $\phi = \cos^{-1} 0.2427 = 75.95°$

$$I_1 = \frac{V}{R} \text{ amp} = \frac{50}{200}$$

hence, $I_1 = 0.25$ A **Ans**

From the phasor diagram, it may be seen that:

$$I_2 = \sqrt{I^2 - I_1^2} \text{ amp} = \sqrt{1.03^2 - 0.25^2}$$

therefore, $I_2 = 1$ A **Ans**

(b) $P = VI \cos \phi \text{ watt} = 50 \times 1.03 \times 0.2427$

so $P = 12.5$ W **Ans**

(c) $X_L = \frac{V}{I_2} \text{ ohm} = \frac{50}{1}$

so, $X_L = 50\ \Omega$

but, $X_L = 2\pi f L$ ohm,

so, $L = \frac{X_L}{2\pi f} \text{ henry} = \frac{50}{2 \times \pi \times 10^3}$

hence, $L = 7.96$ mH **Ans**

Worked Example 2.5

Q   A 1 kΩ resistor is connected in parallel with a pure inductor of 10 mH across a 24 V, 2 kHz supply. Calculate (a) the current through the resistor, (b) the current through the inductor, (c) the current drawn from the supply, and (d) the power dissipated.

A

$R = 1000\,\Omega; L = 0.01\,\text{H}; V = 24\,\text{V}; f = 2000\,\text{Hz}$

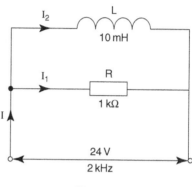

Fig. 2.11

(a) $I_1 = \dfrac{V}{R}$ amp $= \dfrac{24}{1000}$

$I_1 = 24$ mA **Ans**

(b) $X_L = 2\pi fL$ ohm $= 2 \times \pi \times 2000 \times 0.01$

$X_L = 125.7\,\Omega$

$I_2 = \dfrac{V}{X_L}$ amp $= \dfrac{24}{125.7}$

so, $I_2 = 191$ mA **Ans**

(c) $I = \sqrt{I_1^2 + I_2^2}$ amp $= \sqrt{0.024^2 + 0.191^2}$

thus $I = 192.5$ mA **Ans**

(d) $P = I_1^2 R$ watt $= 0.024^2 \times 1000$

and $P = 576$ mW **Ans**

2.4   R-L-C Parallel Circuit

This circuit is a simple extension of the previous two considered. The only differences are that, (a) there are three branch currents, and (b) the circuit phase angle may be either leading or lagging. The latter depends upon the relative values of the currents through inductor and capacitor.

Worked Example 2.6

Q   A 75 Ω resistor, an 80 mH inductor, and a 40 µF capacitor are connected in parallel across an 80 V, 100 Hz supply. Sketch the circuit and phasor diagrams, and calculate (a) the three branch currents, (b) the current drawn from the supply, and (c) the circuit phase angle and power factor.

A

$R = 75\,\Omega; L = 80 \times 10^{-3}\,H; C = 44 \times 10^{-6}\,F; V = 80\,V; f = 100\,Hz$

The circuit and phasor diagrams are shown in Figs. 2.12 and 2.13 respectively.

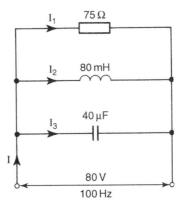

Fig. 2.12

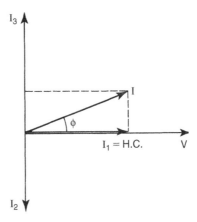

Fig. 2.13

(a) $I_1 = \dfrac{V}{R}\ \text{amp} = \dfrac{80}{75}$

so, $I_1 = 1.067$ A **Ans**

$X_L = 2\pi f L\ \text{ohm} = 2 \times \pi \times 100 \times 80 \times 10^{-3}$

$X_L = 50.265\,\Omega$

$I_2 = \dfrac{V}{X_L}\ \text{amp} = \dfrac{80}{50.265}$

and, $I_2 = 1.592$ A **Ans**

$$X_C = \frac{1}{2\pi f C} \text{ ohm} = \frac{1}{2 \times \pi \times 100 \times 40 \times 10^{-6}}$$

so $X_C = 39.8\ \Omega$

$$I_3 = \frac{V}{X_C} \text{ amp} = \frac{80}{39.8}$$

hence, $I_3 = 2.01$ A **Ans**

(b) From the phasor diagram:

Horizontal component, H.C. $= I_1 = 1.067$ A

Vertical component, V.C. $= I_3 - I_2 = 2.01 - 1.592$

so, V.C. $= 0.418$ A

$$I = \sqrt{\text{V.C.}^2 + \text{H.C.}^2} = \sqrt{0.418^2 + 1.067^2}$$

therefore, $I = 1.146$ A **Ans**

$$\phi = \tan^{-1}\frac{\text{V.C.}}{\text{H.C.}} = \tan^{-1}\frac{0.418}{1.067}$$

hence, $\phi = 21.4°$ leading **Ans**

Power factor, $\cos\phi = 0.931$ leading **Ans**

Worked Example 2.7

Q   An ideal inductor, capacitor and resistor are connected in parallel across a 40 V a.c. supply. The circuit arrangement is shown in Fig. 2.14, and the currents flowing through the inductor is 1.273 A. Calculate (a) the frequency of the supply, (b) the currents through the resistor and capacitor, (c) the current drawn from the supply, and (d) the circuit power factor.

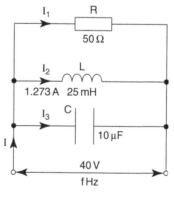

Fig. 2.14

A

$R = 50\ \Omega; L = 0.025\,\text{H}; C = 10^{-5}\,\text{F}; V = 40\text{V}; I_2 = 1.273$ A

(a) Since the three components are connected in parallel then the applied voltage is common to all of them. Also, since the inductor is ideal, then its opposition to current is its inductive reactance only. As inductance is

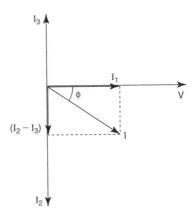

Fig. 2.15

dependent upon the frequency we can use this information to determine the frequency of the supply, thus:

$$X_L = \frac{V}{I_2} \text{ ohm} = \frac{40}{1.273} = 31.42 \ \Omega$$

but, $X_L = 2\pi f 2L$ ohm

$$\text{so, } f = \frac{X_L}{2\pi L} \text{ hertz} = \frac{31.42}{2 \times \pi \times 0.025}$$

thus, $f = 200$ Hz **Ans**

(b)　　$$I_1 = \frac{V}{R} \text{ amp} = \frac{40}{50}$$

so, $I_1 = 0.8$ A **Ans**

$$X_C = \frac{1}{2\pi f C} \text{ ohm} = \frac{1}{2 \times \pi \times 200 \times 10^{-5}}$$

$$X_C = 79.58 \ \Omega$$

$$I_3 = \frac{V}{X_C} \text{ amp} = \frac{40}{79.58}$$

and $I_3 = 0.503$ A **Ans**

(c)　$(I_2 - I_3) = 1.273 - 0.502 = 0.77$ A

$$I = \sqrt{I_1^2 + (I_2 - I_3)^2} \text{ amp} = \sqrt{0.8^2 + 0.77^2}$$

thus, $I = 1.11$ A **Ans**

(d)　　$$\tan \phi = \frac{(I_2 - I_3)}{I_1} = \frac{0.77}{0.8}$$

$$\text{so, } \phi = \tan^{-1} \frac{0.77}{0.8} = \tan^{-1} 0.9625$$

$$\phi = 49.91°$$

$$\text{p.f.} = \cos \phi = \cos 49.91°$$

thus, p.f. $= 0.72$ lagging **Ans**

2.5　Practical Components in Parallel

Thus far, we have considered that the circuit components are perfect: i.e. pure resistor, pure inductor and pure capacitor. In practice, unless the resistor is of the wire-wound type, it is normally assumed to be purely resistive. Similarly, provided that the frequency is not very high, or the capacitor is not an electrolytic type, then the capacitor may also be considered as perfect. The inductor, on the other hand, may rarely be considered as pure inductance. Thus, the resistance of the inductor should be taken into account. The inductor coil therefore has an impedance, Z_{coil} ohm.

A circuit commonly met in practice consists of a 'practical' inductor, connected in parallel with a 'perfect' capacitor. Such a circuit is illustrated in Fig. 2.16, which could represent the coil of an a.c. motor, with a parallel-connected capacitor.

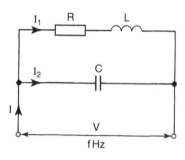

Fig. 2.16

The current through the coil, $I_1 = \dfrac{V}{Z_1}$ amp

where impedance of coil $= Z_1 = \sqrt{R^2 + X_L^2}$ ohm

and $\phi_1 = \cos^{-1}\dfrac{R}{Z_1}$ (lagging V)

Current through the capacitor, $I_2 = V/X_C$ amp and this current will lead the applied voltage by 90°.

The phasor diagram is shown in Fig. 2.17(a). From this diagram it may be seen that the capacitor current forms a vertical component only. The

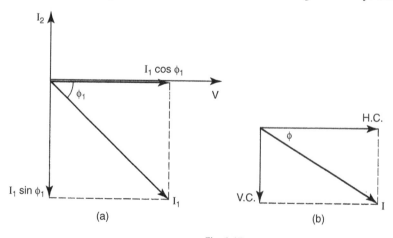

Fig. 2.17

inductor coil current has both horizontal and vertical components, $I_1 \cos \phi_1$ and $I_1 \sin \phi_1$ respectively. The total vertical component (V.C.) is therefore the difference between I_2 and $I_1 \sin \phi_1$. A supplementary phasor diagram is shown in Fig. 2.17(b), from which the circuit current and phase angle can be determined.

Worked Example 2.8

Q   A coil of inductance 159.2 mH and resistance 25 Ω is connected in parallel with a 40 µF capacitor, across a 240 V, 50 Hz supply. Sketch the circuit and phasor diagrams, and calculate (a) the current through, and phase angle of, the coil, (b) the capacitor current, (c) the supply current and phase angle, (d) the power consumed and (e) the circuit impedance.

A

$L = 159.2 \times 10^{-3}\,\text{H}; R = 25\,\Omega; C = 40 \times 10^{-6}\,\text{F}; V = 240\text{V}; f = 50\,\text{Hz}$

The circuit and phasor diagrams are shown in Figs. 2.18 and 2.19 respectively.

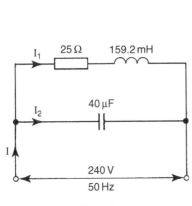

Fig. 2.18

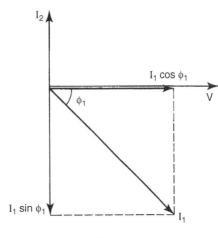

Fig. 2.19

(a) $X_L = 2\pi f L \text{ ohm} = 2 \times \pi \times 50 \times 159.2 \times 10^{-3}$

so, $X_L = 50\,\Omega$

$Z_1 = \sqrt{R^2 + X_L^2} \text{ ohm} = \sqrt{25^2 + 50^2}$

and $Z_1 = 55.9\,\Omega$

$I_1 = \dfrac{V}{Z_1} \text{ amp} = \dfrac{240}{55.9}$

hence, $I_1 = 4.29\,\text{A}$ **Ans**

$\phi_1 = \cos^{-1}\dfrac{R}{Z_1} = \cos^{-1}\dfrac{25}{55.9}$

therefore $\phi = -63.43°$ **Ans**

(b) $X_C = \dfrac{1}{2\pi f C}$ ohm $= \dfrac{1}{2 \times \pi \times 50 \times 40 \times 10^{-6}}$

so, $X_C = 79.58\ \Omega$

$I_2 = \dfrac{V}{X_C}$ amp $= \dfrac{240}{79.58}$

hence, $I_2 = 3.02$ A **Ans** (I_2 leads V by 90°)

(c)   Referring to Fig. 2.20:

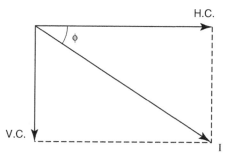

Fig. 2.20

H.C. $= I_1 \cos \phi_1$ amp $= 4.29 \cos 63.43°$

so, H.C. $= 1.919$ A

V.C. $= I_2 - I_1 \sin \phi_1$ amp $= 3.02 - 4.29 \sin 63.43°$

and V.C. $= -0.817$ A

$I = \sqrt{\text{V.C.}^2 + \text{H.C.}^2}$ amp $= \sqrt{-0.817^2 + 1.919^2}$

therefore, $I = 2.09$ A **Ans**

$\phi = \tan^{-1} \dfrac{\text{V.C.}}{\text{H.C.}} = \tan^{-1} \dfrac{-0.817}{1.919}$

hence, $\phi = -23.06°$ **Ans**

(d) $P = VI \cos \phi$ watt $= 240 \times 2.09 \times 0.9201$

so, $P = 461.5$ W **Ans**

Alternatively, $P = I_1^2 R$ watt

(e) $Z = \dfrac{V}{I}$ ohm $= \dfrac{240}{2.09}$

therefore, $Z = 114.8\ \Omega$ **Ans**

Note: In both of the last two examples, the circuit current is *less* than both the current through the coil and that through the capacitor. This would be an impossible situation in the corresponding d.c. circuit, where

$I = I_1 + I_2$ amp. It is possible in the parallel a.c. circuit because I is the PHASOR SUM of the branch currents, and NOT the arithmetic sum.

2.6   Series Resonance

The concept of resonance in a series R-L-C circuit is briefly dealt with in Chapter 1. It will be helpful to review this concept now, and then to study the effect in more detail.

1   The reactances of both a capacitor and an inductor depend on the frequency of the supply.
2   $X_L \propto f$, and $X_C \propto 1/f$.
3   For given values of L and C; at low frequencies X_L will be relatively small, compared to X_C. At high frequencies, X_L will be relatively large, compared to X_C.
4   At one particular frequency, $X_L = X_C$. This is known as the resonant frequency, f_0 hertz.
5   Under resonant conditions, the circuit current and voltage are in phase with each other. The circuit therefore behaves as if it were purely resistive in nature. This is illustrated in the phasor diagram of Fig. 2.21.
6   The circuit impedance Z is at its minimum possible value of R ohm.
7   The resonant frequency,

$$f_0 = \frac{1}{2\pi\sqrt{LC}} \text{ hertz}$$

$$\text{or } \omega_0 = \frac{1}{\sqrt{LC}} \text{ rad/s}$$

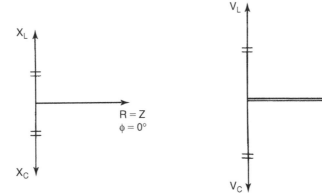

Fig. 2.21

Worked Example 2.9

Q   A 47 pF capacitor is connected in series with a coil of resistance 25 Ω and inductance 50 mH, across a 5 mV variable frequency supply. Calculate (a) the supply frequency that results in resonance, (b) the circuit current, (c) the p.d. across the capacitor and coil.

A

$C = 47 \times 10^{-12}\,\text{F}; \; R = 25\,\Omega; \; L = 50 \times 10^{-3}\,\text{H}; \; V = 5 \times 10^{-3}\,\text{V}$

The circuit and phasor diagrams are shown in Figs. 2.22 and 2.23 respectively.

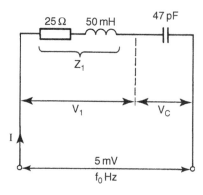

Fig. 2.22

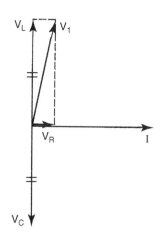

Fig. 2.23

(a) $f_0 = \dfrac{1}{2\pi\sqrt{LC}}$ hertz $= \dfrac{1}{2\pi\sqrt{50 \times 10^{-3} \times 47 \times 10^{-12}}}$

therefore, $f_0 = 103.82$ kHz **Ans**

(b)   At f_0, $Z = R$ ohm $= 25\,\Omega$

and $I = \dfrac{V}{Z}$ amp $= \dfrac{5 \times 10^{-3}}{25}$

hence, $I = 200\,\mu\text{A}$ **Ans**

(c)
$$X_C = \frac{1}{2\pi fC} \text{ ohm} = \frac{1}{2\pi \times 103.82 \times 10^3 \times 47 \times 10^{-12}}$$

so, $X_C = 32.62 \text{ k}\Omega$

$$V_C = IX_C \text{ volt} = 200 \times 10^{-6} \times 32.62 \times 10^3$$

therefore, $V_C = 6.52$ V **Ans**

Since the coil possesses both inductance and resistance, then the total opposition it offers to the current is its impedance, $Z_1 = \sqrt{R^2 + X_L^2}$ ohm. At resonance, $X_L = X_C = 32.62 \text{ k}\Omega$

$$\text{so, } Z_1 = \sqrt{25^2 + (32.62 \times 10^3)^2} = 32.62 \text{ k}\Omega$$

i.e. the coil resistance happens to be negligible compared with its reactance. Thus, the p.d. across the coil is the same as that across the capacitor, namely 6.52 V **Ans**

2.7   Circuit Q-factor

In the previous example, it may be seen that the p.d.s across the capacitor and the inductor are not only equal, but are many times greater than the supply voltage. In this particular case, $V_C = 1305 \times V$.

This effect is known as voltage magnification. The ratio of the p.d. across the capacitor (or inductor) to the applied voltage is called the voltage magnification factor, or Q-factor.

$$\text{Therefore, } Q = \frac{V_C}{V} = \frac{V_L}{V} \tag{2.1}$$

but, $V_C = IX_C$, $V_L = IX_L$, and $V = IR$ $(Z = R)$

therefore, $Q = \dfrac{IX_C}{IR} = \dfrac{IX_L}{IR}$

$$\text{so, } Q = \frac{X_C}{R} = \frac{X_L}{R} \tag{2.2}$$

also, since $X_C = 1/\omega_0 C$, and $X_L = \omega_0 L$ ohm

$$\text{then, } Q = \frac{1}{\omega_0 CR} = \frac{\omega_0 L}{R} \tag{2.3}$$

Both equations (2.2) and (2.3) above require the resonant frequency to be known. However, the Q-factor may also be expressed in terms of the circuit components only, as follows:

$$Q = \frac{1}{\omega_0 CR}, \text{ and } \omega_0 = \frac{1}{\sqrt{LC}}$$

$$\text{therefore, } Q = \frac{\sqrt{LC}}{CR}$$

$$\text{hence, } Q = \frac{1}{R}\sqrt{\frac{L}{C}} \tag{2.4}$$

Worked Example 2.10

Q   A series circuit consists of a $50\,\Omega$ resistor, a $0.15\,H$ inductor and a 22 nF capacitor. This combination is connected across a 24 V a.c. supply. Determine (a) the resonant frequency, (b) the Q-factor, and (c) the p.d. across the capacitor and inductor.

A

$R = 50\,\Omega; L = 0.15\,H; C = 22\times10^{-9}\,F; V = 24V$

(a) $\omega_0 = \dfrac{1}{\sqrt{LC}}$ rad/s $= \dfrac{1}{\sqrt{0.15 \times 22 \times 10^{-9}}}$

so, $\omega_0 = 17.408 \times 10^3$ rad/s

hence, $f_0 = \dfrac{17.408 \times 10^3}{2\pi} = 2.77$ kHz **Ans**

(b) $Q = \dfrac{1}{R}\sqrt{\dfrac{L}{C}} = \dfrac{1}{50}\sqrt{\dfrac{0.15}{22 \times 10^{-9}}}$

so, $Q = 52.22$ **Ans**

(c) $V_C = Q \times V$ volt $= 52.22 \times 24$

hence, $V_C = V_L = 1.253$ kV **Ans**

Worked Example 2.11

Q   An inductor of inductance 0.2 H and resistance $15\,\Omega$ is connected in series with a capacitor across a 10 mV, variable frequency supply. This circuit is required to have a Q-factor of 25. Calculate (a) the value of capacitor required, (b) the resonant frequency, (c) the bandwidth, and (d) the current flowing through this circuit at the resonant frequency.

A

$L = 0.2\,H; R = 15\,\Omega; V = 10^{-2}\,V; Q = 25$

(a) $Q = \dfrac{1}{R}\sqrt{\dfrac{L}{C}}$

$QR = \sqrt{\dfrac{L}{C}}$

$\dfrac{L}{C} = (QR)^2$

and, $C = \dfrac{L}{(QR)^2}$ farad $= \dfrac{0.2}{(25 \times 15)^2}$

so, $C = 1.42\,\mu F$ **Ans**

(b) $$f_0 = \frac{1}{2\pi\sqrt{LC}} \text{ hertz} = \frac{1}{2 \times \pi\sqrt{0.2 \times 1.42 \times 10^{-6}}}$$

thus, $f_0 = 298.6$ Hz **Ans**

(c) $$B = \frac{f_0}{Q} \text{ hertz} = \frac{298.6}{25}$$

so, $B = 11.95$ Hz **Ans**

(d)   In a **series** resonant circuit the only opposition to current flow is the resistance in the circuit, thus:

$$I_0 = \frac{V}{R} \text{ amp} = \frac{10^{-2}}{15}$$
$$\text{thus, } I_0 = 0.667 \text{ mA **Ans**}$$

2.8   Frequency Response Curve

We have seen that, at the resonant frequency, the impedance of the series circuit reaches its minimum value of R ohm. At frequencies greater than and less than f_0, there is an overall reactive component, which results in increased impedance. Thus the current flow through the circuit reaches its maximum possible value at the resonant frequency. A graph of the circuit current versus frequency is known as the frequency response curve, and will have the shape shown in Fig. 2.24. Note that the frequency axis is often plotted to a logarithmic scale (as in the diagram). The main reason for this is to enable a large range of frequencies to be accommodated, without the need for excessively wide graph paper. It also has the effect of making the graph more symmetrical about the vertical axis, on either side of f_0.

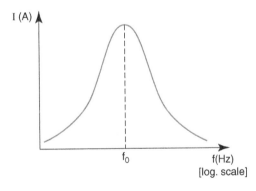

Fig. 2.24

The Q-factor of the circuit will determine the 'peakiness' of the response curve. A high value of Q results in a curve with a high peak and steep sides on either side of f_0. A low value of Q results in a shallow curve, with gently sloping sides. From equations (2.2) to (2.4), for Q-factor, it is apparent that a low value of R gives a high Q, and vice-versa. This effect is illustrated in Fig. 2.25.

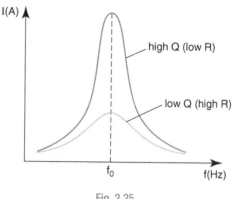

Fig. 2.25

From the curves shown in this diagram, it is apparent that a circuit having a high Q-factor will offer minimal opposition to only a narrow range of frequencies on either side of f_0. This property of the circuit is referred to as its selectivity. The range of frequencies concerned is called the bandwidth of the circuit. In order to make comparisons of selectivity between circuits, the term bandwidth needs to be explained.

Let the current that flows at resonance be referred to as I_0. The bandwidth, B, of the circuit is identified as that range of frequencies over which the circuit current is greater than or equal to $I_0/\sqrt{2}(I \geq I_0/\sqrt{2})$. The current value at the two extremes defined above are shown as I_1 and I_2 in Fig. 2.26. The 'cut-off' frequencies are shown as f_1 and f_2. The circuit bandwidth is therefore the difference between f_2 and f_1.

$$\text{i.e. Bandwidth, } B = (f_2 - f_1) \text{ hertz} \tag{2.5}$$

$$\text{and } I_1 = I_2 = \frac{I_0}{\sqrt{2}} \tag{2.6}$$

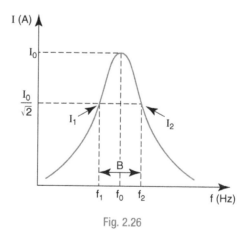

Fig. 2.26

From Fig. 2.25, it should be obvious that the bandwidth for a circuit having a high Q-factor will be less than that for one with a low

Q-factor. From this we can conclude that the bandwidth and Q-factor of a circuit are related to each other. This relationship is as follows:

$$\text{Bandwidth, } B = \frac{f_0}{Q} \text{ hertz} \qquad (2.7)$$

Thus far, we have seen that the Q-factor, and hence the selectivity of the circuit, is inversely proportional to the circuit resistance. Consider now equation (2.4), which is repeated below.

$$Q = \frac{1}{R}\sqrt{\frac{L}{C}}$$

From this equation it may be seen that Q is directly proportional to the ratio L/C. Thus, if the resistance is maintained constant and the ratio of L to C is varied, the Q-factor and selectivity will vary in direct proportion. This is illustrated in Fig. 2.27.

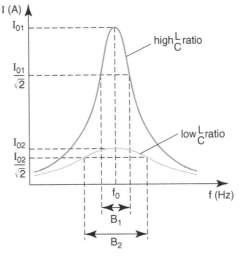

Fig. 2.27

In a practical circuit it is difficult to vary the ratio of L to C without also varying R. If the resistance is kept constant by leaving the coil unchanged, then only C can be varied. However, when this is done, the circuit resonant frequency will also be varied.

Worked Example 2.12

Q   A tuned (resonant) circuit consists of a coil, having 40 Ω resistance and 0.2 H inductance, connected in series with a 22 nF capacitor.

(i) Determine (a) the circuit Q-factor, (b) the resonant frequency, and (c) the circuit bandwidth. (ii) If an additional 40 Ω resistor is added to the circuit, determine the effects on the Q-factor and bandwidth.

A

(i) $R = 40\,\Omega;\ L = 0.2\,\text{H};\ C = 22 \times 10^{-9}\,\text{F}$

(a) $$Q = \frac{1}{R}\sqrt{\frac{L}{C}} = \frac{1}{40}\sqrt{\frac{0.2}{22 \times 10^{-9}}}$$

therefore, $Q = 75.38$ **Ans**

(b) $$f_0 = \frac{1}{2\pi\sqrt{LC}} \text{ hertz} = \frac{1}{2\pi\sqrt{0.2 \times 22 \times 10^{-9}}}$$

so, $f_0 = 2.4$ kHz **Ans**

(c) $$B = \frac{f_0}{Q} \text{ hertz} = \frac{2.4 \times 10^3}{75.38}$$

hence, $B = 31.83$ Hz **Ans**

(ii) $R = 80\,\Omega$; L and C as before

Since R is now twice its original value, then Q must be halved, and B must be doubled. The resonant frequency will remain unchanged at 2.4 kHz. Thus, $Q = 37.69$ and $B = 63.76$ Hz **Ans**

A series circuit, at or near to its resonant condition, offers relatively little opposition to the flow of current through it. It is therefore also known as an acceptor circuit. This effect is utilised in a filter circuit, known as a band-pass filter. Filter circuits are discussed later in this chapter.

2.9 Parallel Resonance

Consider a coil possessing both inductance and resistance, connected in parallel with a capacitor, across an a.c. supply. This circuit is shown in Fig. 2.28. Let the frequency of the supply be the resonant frequency for the circuit. This will mean that the circuit current, I, will be in phase with the supply voltage, V. This condition is illustrated in Fig. 2.29. From this phasor diagram it may be seen that the resonant condition occurs when the two vertical components of current are equal. Thus resonance occurs when $I_1 \sin \phi_1 = I_2$. The result is that the current drawn from the supply, I is equal to $I_1 \cos \phi_1$.

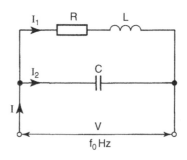

Fig. 2.28

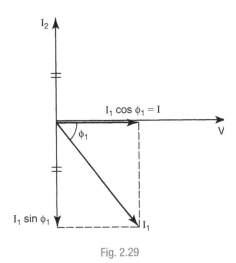

Fig. 2.29

Now, $I_1 = \dfrac{V}{Z_1}$, and considering the impedance triangle for the coil

(Fig. 2.30)

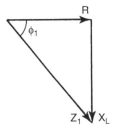

Fig. 2.30

$$\sin \phi_1 = \frac{X_L}{Z_1} = \frac{\omega_0 L}{Z_1}$$

therefore, $I_1 \sin \phi_1 = \dfrac{V \omega_0 L}{Z_1^2}$ ....[1]

also, $I_2 = \dfrac{V}{X_C} = V\omega_0 C$ ......[2]

For resonance, [1] = [2], so

$\dfrac{V\omega_0 L}{Z_1^2} = V\omega_0 C$, and dividing through by $V\omega_0$

$$\frac{L}{Z_1^2} = C$$

so $Z_1^2 = \dfrac{L}{C}$ ....[3]

but, $Z_1 = \sqrt{R^2 + X_L^2}$, so $Z_1^2 = R^2 + X_L^2$

or, $Z_1^2 = R^2 + (\omega_0 L)^2$,

and substituting this into equation [3]:

$$R^2 + \omega_0^2 L^2 = \frac{L}{C}$$

$$\omega_0^2 L^2 = \frac{L}{C} - R^2$$

$$\omega_0^2 = \frac{L}{L^2 C} - \frac{R^2}{L^2} = \frac{1}{LC} - \frac{R^2}{L^2}$$

$$\text{therefore, } \omega_0 = \sqrt{\frac{1}{LC} - \frac{R^2}{L^2}} \text{ rad/s} \qquad (2.8)$$

$$\text{and, } f_0 = \frac{1}{2\pi}\sqrt{\frac{1}{LC} - \frac{R^2}{L^2}} \text{ hertz} \qquad (2.9)$$

Comparing equation (2.9) with that for series resonance, it will be seen that they are very similar. The difference is the inclusion of the term R^2/L^2 in equation (2.9). The significance of this is that, for the same circuit components, the resonant frequency for the parallel circuit is less than that for the series arrangement. However, if the term R^2/L^2 is very much less than $1/LC$, then the equation for parallel resonance *approximates* to:

$$f_0 \simeq \frac{1}{2\pi\sqrt{LC}} \text{ hertz}$$

This condition tends to occur when the resistance of the coil is very small compared with its reactance.

It should be noted that if the supply frequency is either reduced below, or increased above f_0, then the vertical components of current will no longer 'cancel out'. At frequencies other than f_0, the circuit current will therefore be greater. Hence, we can draw the conclusion that at resonance, the impedance of a parallel circuit reaches its MAXIMUM value. This is exactly opposite to the case for a series resonant circuit.

The impedance of the parallel circuit at resonance is referred to as its dynamic impedance, Z_D, or its dynamic resistance, R_D ohm. Since the circuit phase angle at resonance is zero, the circuit again behaves as if it were purely resistive in nature. However, Z_D is NOT EQUAL to the resistance of the coil, R, since this would be the *minimum* possible value. We therefore need to be able to calculate the value of Z_D for a given circuit.

From the phasor diagram (Fig. 2.29), and the impedance triangle (Fig. 2.30), we have:

$$\tan \phi_1 = \frac{\omega_0 L}{R} = \frac{I_1 \sin \phi_1}{I}$$

$$\text{but, } I = \frac{V}{Z_D}$$

$$\text{so } \frac{\omega_0 L}{R} = \frac{I_1 \sin \phi_1 Z_D}{V}$$

also, $I_1 \sin \phi_1 = I_2$

therefore, $\dfrac{\omega_0 L}{R} = \dfrac{I_2}{V} Z_D$; and $\dfrac{I_2}{V} = \omega_0 C$

so $\dfrac{\omega_0 L}{R} = \omega_0 C Z_D$; and dividing through by ω_0:

$$\frac{L}{R} = CZ_D$$

hence, $Z_D = \dfrac{L}{CR}$ ohm   (2.10)

Worked Example 2.13

Q   A coil of inductance 15 mH and resistance 50 Ω, is connected in parallel with a 1 μF capacitor, across a 20 V variable frequency a.c. supply. Determine (a) the approximate and actual values for the resonant frequency, (b) the dynamic impedance, (c) the current drawn from the supply, under resonant conditions, and (d) the corresponding capacitor current.

A

$L = 15 \times 10^{-3}$ H; $R = 50\,\Omega$; $C = 10^{-6}$ F; $V = 20$V

(a) $f_0 = \dfrac{1}{2\pi\sqrt{LC}}$ hertz $= \dfrac{1}{2\pi\sqrt{15 \times 10^{-3} \times 10^{-6}}}$

so $f_0 = 1.3$ kHz **Ans** (approximate value)

$$f_0 = \frac{1}{2\pi}\sqrt{\frac{1}{LC} - \frac{R^2}{L^2}} \text{ hertz}$$

$$= \frac{1}{2\pi}\sqrt{\frac{1}{15 \times 10^{-3} \times 10^{-6}} - \frac{50^2}{(15 \times 10^{3})^2}}$$

$$= \frac{1}{2\pi}\sqrt{6.67 \times 10^7 - 1.11 \times 10^7}$$

so, $f_0 = 1.19$ kHz **Ans** (actual value)

(b) $Z_D = \dfrac{L}{CR}$ ohm $= \dfrac{15 \times 10^{-3}}{50 \times 10^{-6}}$

$Z_D = 300\,\Omega$ **Ans**

(c) $I = \dfrac{V}{Z_D} = \dfrac{20}{300}$

hence, $I = 66.7$ mA **Ans**

(d) $I_C = \dfrac{V}{X_C}$ amp; where $X_C = \dfrac{1}{2\pi f_0 C}$ ohm

so $I_C = V \times 2\pi f_0 C = 20 \times 2\pi \times 1.19 \times 10^3 \times 10^{-6}$

and, $I_C = 149.5$ mA **Ans**

Worked Example 2.14

Q   A coil of resistance $20\,\Omega$ and inductance $10\,\text{mH}$ is connected in parallel with a $2.2\,\mu\text{F}$ capacitor across a $15\,\text{V}$ variable frequency supply. Calculate (a) the frequency at which the current drawn from the supply is at its minimum value, (b) the value of this current, (c) the circuit Q-factor and bandwidth, (d) the lower and upper cut-off frequencies, and (e) the current circulating between inductor and capacitor.

A

$R = 20\,\Omega; L = 10^{-2}\,\text{H}; C = 2.2 \times 10^{-6}\,\text{F}; V = 15\,\text{V}$

(a) For the circuit current to be at its minimum value, the supply frequency must be the resonant frequency for the circuit. Hence:

$$f_0 = \frac{1}{2\pi}\sqrt{\frac{1}{LC} - \frac{R^2}{L^2}}\ \text{hertz} = \frac{1}{2\pi}\sqrt{\frac{1}{10^{-2} \times 2.2 \times 10^{-6}} - \frac{400}{10^{-4}}}$$

$$= \frac{1}{2\pi}\sqrt{45.45 \times 10^6 - 4 \times 10^6} = \frac{1}{2\pi}\sqrt{41.45 \times 10^6}$$

thus $f_0 = 1.025\,\text{kHz}$ **Ans**

(b) For a parallel resonant circuit, the opposition offered to the supply current is the circuit dynamic impedance, Z_D, where,

$$Z_D = \frac{L}{CR}\ \text{ohm} = \frac{10^{-2}}{2.2 \times 10^{-6} \times 20}$$

so $Z_D = 227.27\,\Omega$

and $I_0 = \dfrac{V}{Z_D}\ \text{amp} = \dfrac{15}{227.27}$

thus $I_0 = 66\,\text{mA}$ **Ans**

(c)
$$Q = \frac{1}{R}\sqrt{\frac{L}{C}} = \frac{1}{20}\sqrt{\frac{10^{-2}}{2.2 \times 10^{-6}}}$$

and $Q = 3.37$ **Ans**

$$B = \frac{f_0}{Q}\ \text{hertz} = \frac{1025}{3.37}$$

so, $B = 304\,\text{Hz}$ **Ans**

(d) The lower cut-off frequency, $f_1 = f_0 - \dfrac{B}{2}$ hertz (see Fig. 2.31)

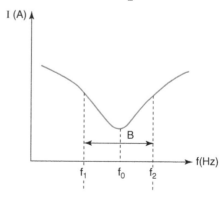

Fig. 2.31

$$f_1 = 1025 - 304/2$$
so $f_1 = 873$ Hz **Ans**

Upper cut-off frequency, $f_2 = f_0 + B/2$ hertz

so $f_2 = 1025 + 304/2$
and $f_2 = 1.177$ kHz **Ans**

(e) In a parallel resonant circuit the current circulating between capacitor and inductor will be Q times the supply current, thus

$$I_C = QI_0 \text{ amp} = 3.37 \times 66 \times 10^{-3}$$
and $I_C = 222$ mA **Ans**

2.10 The Importance of Power Factor

In situations where large amounts of power are generated and consumed, the power factor of the load being supplied becomes increasingly important. It is possible for the load, and its power factor, to change. For this reason, most a.c. electrical machines, such as alternators (generators) and transformers are rated in VA or kVA. The kVA rating of such a machine indicates the highest current that it can safely supply, at its rated output voltage. For example, a 350 V alternator rated at 525 kVA can supply a maximum possible current of 1500 A. The *power* output of the alternator would be 525 kW *only* if the load was purely resistive i.e. if the power factor was unity $(\cos \phi = \cos 0° = 1)$. For any power factor less than 1, the power output would be less than the kVA figure. This is of course the normal situation in practice.

Consider now, the above specified alternator, supplying a load whose power factor can be switched between (say) 1 and 0.5. The turbine driving the alternator must develop 525 kW plus the power losses in the alternator, in order to yield 525 kVA at the alternator output. If the load power factor is unity, then the load will develop a power of 525 kW, at a current of 1500 A. If the load power factor is 0.5, then it will dissipate only 262.5 kW, at a current of 1500 A, since $P = VI \cos \phi$ watt. The alternator however, still has to provide the same voltage and current as before. The load is therefore not making efficient use of the available power from the supply. This situation may be summed up by saying that, *for a given power*, the lower the load power factor, the larger must be the alternator to generate that power. The result is inefficient usage of the supply, and an increase of costs. For this reason the supply authorities always try to attempt to improve the power factor of their loads. They also encourage large industrial consumers to do likewise, by an additional tariff imposed on the reactive component (kVAr) supplied.

2.11   Power Factor Correction

We have seen that when a capacitor is connected in parallel with an inductive circuit unity power factor is obtained, at the resonant frequency. Alternatively, the resonant condition can be achieved by altering the value of the capacitor, so that the resonant and supply frequencies are one and the same. The vast majority of practical loads are inductive in nature, resulting in a power factor of less than 1. In these cases, the power factor may be improved (increased) by a parallel connected capacitor. Consider the following example.

Worked Example 2.15

Q   A motor draws a current of 3.5 A at a lagging power factor of 0.6, when connected to a 240 V, 50 Hz supply. Calculate (a) the value of parallel connected capacitor that will correct this power factor to unity, and (b) the value of current now drawn from the supply.

A

$I_1 = 3.5$ A; $\cos \phi_1 = 0.6$; $V = 240$ V; $f = 50$ Hz

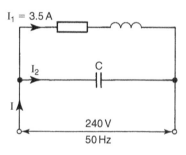

Fig. 2.32

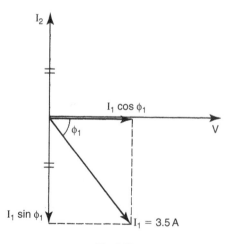

Fig. 2.33

(a) The appropriate circuit and phasor diagrams are shown in Figs. 2.32 and 2.33 respectively.

Thus, $I_2 = I_1 \sin \phi_1$ (see Fig. 2.33)

hence $I_2 = 3.5 \times 0.8 = 2.8$ A

$$X_C = \frac{V}{I_2} \text{ ohm} = \frac{240}{2.8}$$

therefore, $X_C = 85.71 \ \Omega$

$$X_C = \frac{1}{2\pi f C}; \text{ so } C = \frac{1}{2\pi f X_C} \text{ farad}$$

hence $C = \dfrac{1}{2 \times \pi \times 50 \times 85.71}$

therefore, $C = 37.14 \ \mu\text{F}$ **Ans**

$\cos \phi_1 = 0.6;$ so $\phi_1 = 53.13°$ and $\sin \phi_1 = 0.8$

For unity power factor, the supply current must be in phase with the supply voltage.

(b) From Fig. 2.33, it may be seen that $I = I_1 \cos \phi_1$

so, $I = 3.5 \times 0.6$

hence, $I = 2.1$ A **Ans**

The following points should be noted. Firstly, by improving the overall load power factor as above, the current drawn from the supply has been considerably reduced (by 40%), yet the load power dissipation remains unchanged. However, it is not normal practice to correct the power factor to unity, because this results in the resonant condition. This is avoided in power circuits, since the current circulating between the capacitor and inductor can be many times the supply current. For this reason, in practical circuits, the power factor is not normally improved beyond 0.9. In addition, the closer we get to $\cos \phi = 1$, the larger the capacitor value required. Large value capacitors having a high working voltage are relatively expensive components. Consider the following example.

Worked Example 2.16

Q For the motor specified in the previous example (2.15), calculate (a) the value of capacitor required to improve the load power factor to 0.9, and (b) the current now drawn from the supply.

A

In order to solve this problem a phasor diagram is an important requirement. This phasor diagram is shown in Fig. 2.34.

(a) From the phasor diagram:

$$I_1 \cos \phi_1 = 3.5 \times 0.6 = 2.1 \text{ A}$$

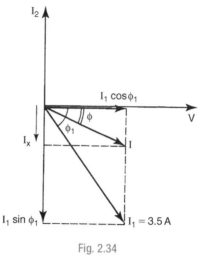

Fig. 2.34

Also, considering the triangle consisting of I, $I_1 \cos \phi_1$ and I_x; where $\cos \phi = 0.9$:

$$\phi = \cos^{-1} 0.9 = 25.84°; \text{ so } \tan \phi = 0.4843$$

but $\tan \phi = \dfrac{I_x}{I_1 \cos \phi_1}$; so $I_x = I_1 \cos \phi_1 \times \tan \phi$

therefore, $I_x = 2.1 \times 0.4843 = 1.017$ A

Now, $I_2 = I_1 \sin \phi_1 - I_x$

so $I_2 = (3.5 \times 0.8) - 1.017 = 1.783$ A

$$X_C = \frac{V}{I_2} \text{ ohm} = \frac{240}{1.783} = 134.61 \; \Omega$$

and $C = \dfrac{1}{2\pi f X_C}$ farad $= \dfrac{1}{2 \times \pi \times 50 \times 134.61}$

hence $C = 23.65 \; \mu$F **Ans**

(b) Once more, from the phasor diagram it may be seen that:

$$\cos \phi = \frac{I_1 \cos \phi_1}{I} \text{ so } I = \frac{I_1 \cos \phi_1}{\cos \phi}$$

hence $I = \dfrac{2.1}{0.9} = 2.33$ A **Ans**

From this example, it is clear that a smaller (and probably cheaper) capacitor is required, and the current drawn from the supply is still reduced considerably (by 33.3%). This example also illustrates the importance and usefulness of a phasor diagram.

2.12   Filters

A filter is a network designed to pass signals at certain frequencies, and to reject signals at all other frequencies. Ideally, a filter would introduce zero attenuation to selected frequencies, known as the pass band(s), and completely block all others (the stop band). Since a filter has to be frequency sensitive, then it must employ frequency dependent components; i.e. inductors or capacitors, or (usually) a combination of both of these components. There are four basic types of filter, namely, low-pass, high-pass, band-pass and band-stop. In addition, a filter circuit may be classified as being either passive or active. A passive filter is one which contains only a combination of resistors, inductors or capacitors, but no power source within it. An active filter consists of a combination of these components interconnected with an operational amplifier (op amp). The latter would include its own power supply source, independent of the signal to be filtered.

2.13   Low-pass Filters

A very simple form of low-pass filter is shown in Fig. 2.35. The input voltage is shown as V_1 and the output voltage as V_0. Since the reactance of the capacitor, $X_C \propto 1/f$, then at low frequencies, this reactance will be very high. Provided that the value of X_C is very much greater than the value of R, then V_0 will be almost the same value as V_1. The reason is that the circuit forms a potential divider circuit, such that $V_0 = [X_C/(R + X_C)] \times V_1$. As the frequency of the input is increased, so the value of X_C will decrease, but R remains constant. Thus, as f increases, so V_0 decreases. The frequency response curve for this simple arrangement is illustrated in Fig. 2.36. In this diagram, the vertical axis is the *ratio* of the output voltage to the input voltage (V_0/V_1). The ideal response is shown by the dotted lines. The 'cut-off' frequency is shown as f_c.

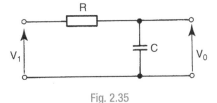

Fig. 2.35

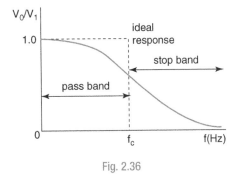

Fig. 2.36

So far, we have considered the behaviour of the circuit in terms of the input and output voltages. A similar argument applies if we consider the ratio of the input and output currents. Since at low frequencies

X_C is high, virtually no current flows through the capacitor. Thus, the current reaching the output terminals is virtually the same as that entering at the input. At high frequencies, X_C is very low. In this case, the capacitor will divert (shunt) most of the current, thus preventing it from reaching the output terminals.

One problem with this form of circuit is the inclusion of the resistor. This will result in some unwanted attenuation of the signal in the low frequency pass band. One method of overcoming this problem is to use an active form of filter. In this case, the amplifier employed can make up for any *unwanted* attenuation. You are not required to explain the full circuit action of active filters, but a simplified circuit is shown in Fig. 2.37. The general block diagram symbol for a low-pass filter is shown in Fig. 2.38.

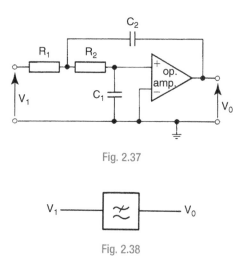

Fig. 2.37

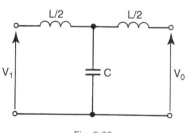

Fig. 2.38

Another form of passive low-pass filter utilises capacitors and inductors. In order to minimise any unwanted attenuation, the inductors need to have as small a resistance as possible. These filters may be in the form of either a 'T' or a 'π' network. In each case, the series connected elements are the inductors, and the parallel elements are the capacitors. An example of a 'T' network low-pass filter appears in Fig. 2.39. The 'π' type is shown in Fig. 2.40.

Fig. 2.39

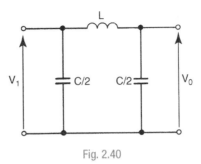

Fig. 2.40

The action of both types is the same. At low frequencies, the reactance of the inductors (X_L) is low, since $X_L \propto f$. On the other hand, at low frequencies, the reactance of the capacitors will be very high. Hence, at these frequencies, the inductors offer little opposition to the signal, and the capacitors have negligible shunting effect. At high frequencies, the inductive reactance provides considerable attenuation of the signal. At the same time, the capacitors have considerable shunting effect. The frequency response curve for this of type of filter is closer to the ideal than that of the simple *R-C* filter, and is illustrated in Fig. 2.41. One common example of the use of an *L-C* filter is the smoothing circuit employed between a full-wave bridge rectifier and the d.c. load being supplied. In this application, the filter is required to reduce the ripple in the waveform, thus resulting in a smooth d.c.

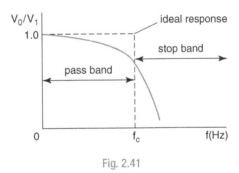

Fig. 2.41

Note: The total inductance of inductors in series is given by their sum. Similarly, the total capacitance of two capacitors in parallel is given by their sum. Thus, in the circuits of Figs. 2.39 and 2.40, the total inductance and capacitance is *L* henry and *C* farads respectively.

2.14   High-pass Filters

The action of this filter is exactly the opposite to that of a low-pass filter. Ideally, it will stop all signals at frequencies below the cut-off value, and pass all frequencies above this value. A simple high-pass filter may also be achieved by the use of a *C-R* circuit, as shown in

Fig. 2.42. At low frequencies, the high reactance of the capacitor severely attenuates the signal, and only a very small proportion of the input voltage is developed (as the output) across the resistor.

At high frequencies, the capacitive reactance is very low, and the output developed across the resistor is relatively large. The frequency response curve for this simple form of filter is illustrated in Fig. 2.43. Figures 2.44 and 2.45 show an active version of this filter, and the block diagram symbol for a high-pass filter, respectively.

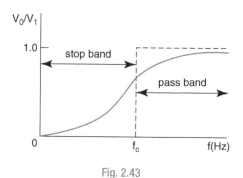

Fig. 2.42

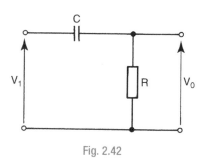

Fig. 2.43

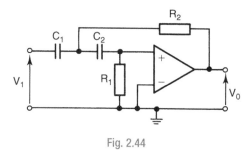

Fig. 2.44

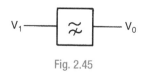

Fig. 2.45

This simple form of *C-R* filter suffers from the same disadvantages as its low-pass version. Thus it is more common to use the *L-C* types, in either 'T' or 'π' configuration. These are shown in Figs. 2.46 and 2.47

with the frequency response curve illustrated in Fig. 2.48. It is left for the reader to confirm the action of the circuit, bearing in mind that the positions of capacitors and inductors in the circuit are reversed when compared with the corresponding low-pass versions.

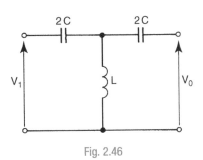

Fig. 2.46

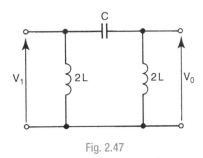

Fig. 2.47

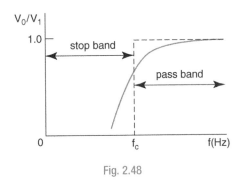

Fig. 2.48

2.15   Band-pass Filters

This circuit is a combination of the *L-C* forms of the previous two types of filter circuits. It utilises both series and parallel tuned (resonant) circuits. The capacitor and inductor values are chosen so that the same resonant frequency applies to both the series and parallel arrangements. The series impedance of the circuit is provided by two identical series tuned circuits. The shunt impedance is provided by the parallel tuned circuit, as shown in Fig. 2.49.

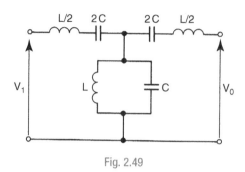

Fig. 2.49

For signals at or near to f_0, the series impedance will be at its minimum possible value. The impedance of the parallel branch will be at its maximum value under these conditions. Thus, over a narrow band of frequencies, the opposition to the signal flow between input and output is minimal. At the same time, the shunting effect on the signal is also minimal. The attenuating effect of the whole circuit is therefore at its minimum.

At frequencies on either side of resonance, the series impedance is increased, whilst the shunt impedance is decreased. Hence, the signal flow between input and output is impeded, and the shunting effect is increased. The signal is thus attenuated as illustrated in Fig. 2.50. The selectivity, and hence bandwidth, of the circuit is determined by the ratio L/C. Figure 2.51 shows the block diagram symbol for a band-pass filter.

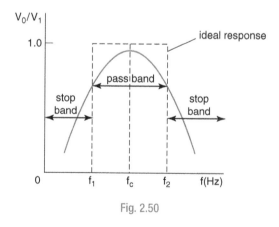

Fig. 2.50

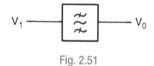

Fig. 2.51

2.16   Band-stop (Notch) Filters

This filter circuit also utilises both series and parallel tuned elements. Compared with the band-pass filter, the roles of the series and parallel tuned elements are reversed. Thus, the series impedance is provided by two identical parallel-tuned elements. The shunt impedance is provided by a series-tuned element. Once more, the components are chosen so that a common resonant frequency is achieved. The circuit arrangement is shown in Fig. 2.52, and its action is as follows.

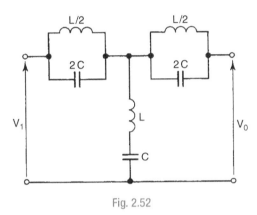

Fig. 2.52

At or near to f_0, the series impedance will be at its maximum, and the shunt impedance will be at its minimum. Under these conditions maximum attenuation is achieved. At frequencies removed from resonance the attenuating effect of the circuit is reduced. This results in a frequency response curve as shown in Fig. 2.53. The band-stop filter symbol is as shown in Fig. 2.54.

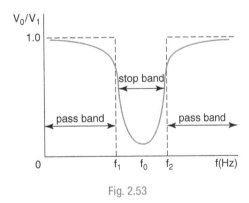

Fig. 2.53

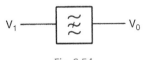

Fig. 2.54

2.17   Coupled Tuned Circuits

In communication systems it would be ideal to have a circuit having a frequency response curve that has a flat top and vertical sides. Such a response curve is shown in Fig. 2.50, as the ideal response for a band-pass filter. This would allow an amplifier to be designed having a large (and uniform) gain over the required bandwidth, and zero gain over all other frequencies. A close approximation to this ideal curve may be achieved by the use of coupled tuned circuits.

A coupled tuned circuit is obtained by having two circuits of the same resonant frequency, and weakly coupling them by a common element; e.g. mutual inductance, self inductance or capacitance. Let us consider one example of this, where the coupling is provided by mutual inductance. A typical circuit is shown in Fig. 2.55. This arrangement is also known as transformer coupling. In this case, the degree of coupling between the two coils depends upon their coupling coefficient k, where $k = M/\sqrt{L_1 L_2}$ (see *Fundamental Electrical and Electronic Principles*, Chapter 5). The overall shape of the resulting frequency response curve varies according to the degree of coupling achieved, and is illustrated in Fig. 2.56.

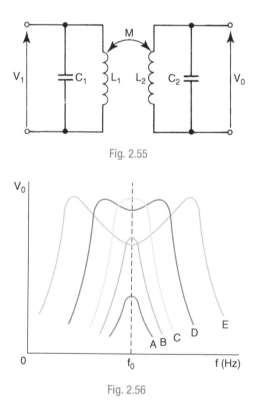

Fig. 2.55

Fig. 2.56

Curves A and B are the result of very loose coupling. This may be defined as the case where the coupling coefficient is less than the reciprocal of the circuit Q-factor, i.e. $k < 1/Q$. Since the coupling is very loose, the impedance and response curve for each individual tuned circuit is virtually unaffected by the presence of the other. The overall response curve is given by the product of the two individual responses. Thus, the output voltage tends to be small, but the selectivity is high.

Curve C shows the overall response when critical coupling is achieved $(k = 1/Q)$. This condition results in a flat top and steep sides for the response curve. In this case, the output at the centre frequency is larger than in any other. The critically coupled circuit therefore approaches the ideal shape required.

If the coupling is increased beyond the critical value, so that $k > 1/Q$, then we have overcoupling. Under this condition the overall response curve splits into two separate peaks, as shown in curves D and E. The greater the coupling the wider the spacing of the peaks and the lower the dip between them.

Summary of Equations

Series resonance: Resonant frequency, $f_0 = \dfrac{1}{2\pi\sqrt{LC}}$ hertz

Circuit Q-factor: $Q = \dfrac{1}{R}\sqrt{\dfrac{L}{C}}$

Bandwidth: $B = \dfrac{f_0}{Q}$ hertz

Parallel resonance: Resonant frequency, $f_0 = \dfrac{1}{2\pi}\sqrt{\dfrac{1}{LC} - \dfrac{R^2}{L^2}}$ hertz

Dynamic impedance, $Z_D = \dfrac{L}{CR}$ ohm

Assignment Questions

1   A $100\,\Omega$ resistor is connected in parallel with a $10.61\,\mu F$ capacitor, across a $120\,V$, $200\,Hz$ supply. Determine (a) the current through each component, (b) the current drawn from the supply, and (c) the circuit impedance.

2   A $20\,\Omega$ resistor is connected in parallel with an inductor of negligible resistance, and inductance of $2.5\,mH$, across a $50\,V$, $1\,kHz$ supply. Determine (a) the current in each branch, (b) the supply current, (c) the circuit impedance and, (d) the power consumed.

3   A coil of negligible resistance and inductance L henry is connected in parallel with a $50\,\Omega$ resistor, across a $250\,V$, $50\,Hz$ supply. If the supply current is $10\,A$, determine (a) the current drawn by each component, (b) the value of the inductance, and (c) the circuit phase angle and power factor.

4   A $1.5\,\mu F$ capacitor is connected in parallel with a $15\,\Omega$ resistor across a $12\,V$, $10\,kHz$ supply. Calculate (a) the current in each branch, (b) the supply current, (c) the circuit impedance, (d) the power consumed, and (e) the values of apparent and reactive components of power.

5   A capacitor of C farads is connected in parallel with resistor of R ohms across a $40\,V$, $150\,Hz$ supply. If the supply current is $0.5\,A$ at a power factor of 0.8 leading, calculate the values of C and R.

6   A $75\,mH$ inductor is connected in parallel with a $10\,\mu F$ capacitor across a $50\,V$, $200\,Hz$ supply. Determine (a) the branch currents, (b) the supply current, (c) the circuit impedance, and (d) the power dissipated.

7   A coil of resistance $50\,\Omega$ and inductance $320\,mH$ is connected in parallel with a $15\,\mu F$ capacitor, across a $250\,V$, $50\,Hz$ supply. Calculate (a) the coil current, (b) the capacitor current, (c) the supply current, (d) the circuit phase angle and impedance, (e) the power consumed, and (f) the apparent and reactive components of power.

8   A coil of resistance $60\,\Omega$ and inductance $0.5\,H$ is connected in parallel with a capacitor, across a $200\,V$, $50\,Hz$ supply. If the capacitor draws a current of $942.5\,mA$, calculate (a) the capacitor value, (b) the current through the coil, (c) the supply current, (d) the circuit phase angle, and (e) the power consumed.

9   A coil of inductance $0.15\,H$ and resistance $30\,\Omega$ is connected in series with a $47\,nF$ capacitor, across a $200\,mV$, variable frequency supply. Determine (a) the frequency at which resonance occurs, (b) the circuit Q-factor, (c) the capacitor p.d. at resonance, and (d) the circuit current.

10   A $50\,mH$ coil of resistance $15\,\Omega$ is connected in series with a variable capacitor, across a $6\,V$, $500\,Hz$ supply. Calculate (a) the capacitor value that results in unity power factor for the circuit, (b) the resulting current, (c) the p.d. across the capacitor, and (d) the p.d. across the coil.

11   A $500\,nF$ capacitor is connected in series with a coil, across a $5\,V$, $1.5\,kHz$ supply. Under these conditions, the current has a value of $50\,mA$, and is in phase with the applied voltage. Determine (a) the resistance and inductance of the coil, (b) the circuit Q-factor, and (c) the circuit bandwidth.

12   A $5\,\mu F$ capacitor is connected in parallel with a $10\,mH$ inductor of resistance $2.5\,\Omega$. Calculate (a) the resonant frequency, (b) the resonant frequency if an additional $22\,\Omega$ resistor is connected in series with the coil, and (c) the Q-factor in each case.

13   A coil of inductance $100\,mH$ and resistance $400\,\Omega$ is connected in parallel with a $10\,nF$ capacitor, across a $12\,V$ a.c. supply. Determine, for the condition when the supply current is a minimum, (a) the frequency, (b) the circuit impedance, (c) the supply current, and (d) the Q-factor.

14   A capacitor is connected in parallel with a coil of inductance $300\,mH$ and resistance $10\,\Omega$. Resonance occurs when the supply is $40\,V$ at a frequency of $2\,kHz$. Calculate, (a) the capacitor value, (b) the supply current, (c) the circuit Q-factor, and (d) the circuit bandwidth.

15   A $5\,kW$ motor has a power factor of 0.6 lagging. When it is connected to a $240\,V$, $50\,Hz$ supply, determine (a) the current drawn from the supply, (b) the value of power factor correction capacitor to produce unity power factor, (c) the supply current under this condition. Explain why unity power factor would normally be avoided for this type of circuit.

16   For the motor and supply specified in Question 15 above, calculate (a) the value of

Assignment Questions

capacitor required to improve the original power factor to 0.9 lagging, (b) this capacitor's VAr rating, and (c) the current now drawn from the supply.

17 A 240V, 50 Hz supply feeds the following loads: (i) incandescent lamps taking a current of 6.5 A at unity power factor; (ii) fluorescent lamps taking a current of 8 A at a power factor of 0.9 leading; and (iii) a motor taking a current of 15 A at a lagging power factor of 0.625. Determine (a) the total current drawn from the supply, and the overall power factor, and (b) the value of power factor correction capacitor required to improve the overall power factor to unity.

Suggested Practical Assignments

Note: Component values and specific items of equipment when quoted here are only suggestions. Those used in practice will naturally depend upon availability within a given institution.

Assignment 1

To investigate the effect of parallel resonance.

Apparatus:

1 × signal generator
1 × inductor (of known value)
1 × capacitor (of known value)
1 × voltmeter
1 × ammeter

Method:

1   Connect the circuit of Fig. 2.57, and calculate the theoretical resonant frequency (f_0) of the circuit.

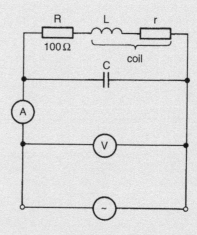

Fig. 2.57

2   Set the signal generator frequency to $f_0/10$, and record this, together with the values for the applied voltage and circuit current.

3   Ensuring that the applied voltage is maintained constant throughout, increase the frequency, in discrete steps, up to $f_0 \times 10$. Record the current at each step.

Note: Take 'extra' readings at, and on either side of the resonant frequency.

4   From your recorded values, calculate and tabulate the corresponding values of circuit impedance.

5   Plot (on the same axes) graphs of circuit current and impedance, versus frequency.

6   From your graphs, determine the actual resonant frequency of the circuit, and also determine the resistance value of the inductor.

7   Complete an assignment report. This should include the circuit diagram, a description of the procedure carried out, all tabulated readings, calculations, and conclusions drawn.

Assignment 2

To observe the effects of power factor correction.

Apparatus:

$1 \times 0.1\,H$ coil
$1 \times$ variable capacitor, 0.1 to $10\,\mu F$
$1 \times 100\,\Omega$ resistor
$1 \times$ double-beam oscilloscope
$1 \times$ ammeter
$1 \times$ a.c. signal generator

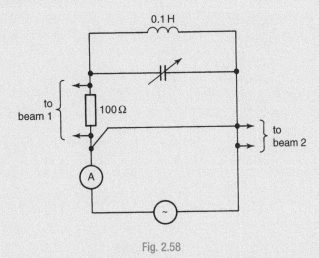

Fig. 2.58

Method:

1 Connect the circuit as in Fig. 2.58.
2 Set the signal generator to 20 V, at a frequency of 500 Hz.
3 Disconnect the capacitor, and measure the circuit current. Note the phase angle of the circuit by comparing the two waveforms on the oscilloscope.
4 Set the capacitor to 0.1 μF, and reconnect it to the circuit.
5 Measure the circuit current and monitor the phase angle on the oscilloscope.
6 Repeat the procedure of paragraph (5) above, for increasing values of capacitance.
7 Complete an assignment report.

Three-Phase A.C. Circuits

Learning Outcomes

This chapter introduces the concepts and principles of the three-phase electrical supply, and the corresponding circuits. On completion you should be able to:

1 Describe the reasons for, and the generation of the three-phase supply.
2 Distinguish between star (3 and 4-wire) and delta connections.
3 State the relative advantages of three-phase systems compared with single-phase-systems.
4 Solve three-phase circuits in terms of phase and line quantities, and the power developed in three-phase balanced loads.
5 Measure power dissipation in both balanced and unbalanced three-phase loads, using the 1, 2 and 3-wattmeter methods, and hence determine load power factor.
6 Calculate the neutral current in a simple unbalanced 4-wire system.

3.1   Generation of a Three-Phase Supply

In order to understand the reasons for, and the method of generating a three-phase supply, let us firstly consider the generation of a single-phase supply. Alternating voltage is provided by an a.c. generator, more commonly called an *alternator*. The basic principle was outlined in *Fundamental Electrical and Electronic Principles,* Chapter 5. It was shown that when a coil of wire, wound on to a rectangular former, is rotated in a magnetic field, an alternating (sinusoidal) voltage is induced into the coil. You should also be aware that for electromagnetic induction to take place, it is the *relative* movement between conductor and magnetic flux that matters. Thus, it matters not whether the field is static and the conductor moves, or vice versa.

For a practical alternator it is found to be more convenient to rotate the magnetic field, and to keep the conductors (coil or winding) stationary.

In any rotating a.c. machine, the rotating part is called the rotor, and the stationary part is called the stator. Thus, in an alternator, the **field** system is contained in the rotor. The winding in which the emf is generated is contained in the stator. The reasons for this are as follows:

> In this context, the term **field** refers to the magnetic field. This field is normally produced by passing d.c. current through the rotor winding. Since the winding is rotated, the current is passed to it via copper slip-rings on the shaft. The external d.c. supply is connected to the slip-rings by a pair of carbon brushes.

(a) When large voltages are generated, heavy insulation is necessary. If this extra mass has to be rotated, the driving device has to develop extra power. This will then reduce the overall efficiency of the machine. Incorporating the winding in the stator allows the insulation to be as heavy as necessary, without adversely affecting the efficiency.

(b) The contact resistance between the brushes and slip-rings is very small. However, if the alternator provided high current output (in hundreds of ampere), the I^2R power loss would be significant. The d.c. current (excitation current) for the field system is normally only a few amps or tens of amps. Thus, supplying the field current via the slip-rings produces minimal power loss. The stator winding is simply connected to terminals on the outside of the stator casing.

(c) For very small alternators, the rotor would contain permanent magnets to provide the rotating field system. This then altogether eliminates the need for any slip-rings. This arrangement is referred to as a brushless machine.

The basic construction for a single-phase alternator is illustrated in Fig. 3.1. The conductors of the stator winding are placed in slots

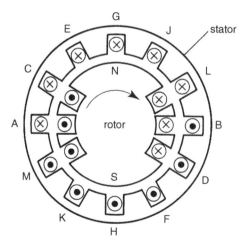

Fig. 3.1

around the inner periphery of the stator. The two ends of this winding
are then led out to a terminal block on the casing. The rotor winding
is also mounted in slots, around the circumference of the rotor. This
figure is used to illustrate the principle. A practical machine would
have many more conductors and slots.

Since the conductors of the stator winding are spread around the whole
of the slots, it is known as a distributed winding. As the rotor field
sweeps past these conductors an emf is induced in each of them in turn.
These individual emfs reach their maximum values only at the instant
that the rotating field 'cuts' them at 90°. Also, since the slots have an
angular displacement between them, then the conductor emfs will be
out of phase with each other by this same angle. In Fig. 3.1 there are a
total of twelve conductors, so this phase difference must be 30°. The
total stator winding emf will therefore not be the *arithmetic* sum of the
conductor emfs, but will be the *phasor* sum, as shown in Fig. 3.2. The
ratio of the phasor sum to the arithmetic sum is called the distribution
factor. For the case shown (a fully distributed winding) the distribution
factor is 0.644.

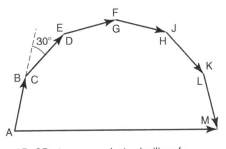

AB, CD etc. are conductor (coil) emfs.
AM is the phasor sum

Fig. 3.2

Now, if all of the stator conductors could be placed into a single pair
of slots, opposite to each other, then the induced emfs would all be
in phase. Hence the phasor and arithmetic sums would be the same,
yielding a distribution factor of unity. This is not a practical solution.
However, if the conductors are concentrated so as to occupy only one
third of the available stator slots, then the distribution factor becomes
0.966. In a practical single-phase alternator, the stator winding is
distributed over two thirds of the slots.

Let us return to the option of using only one third of the slots. We will
now have the space to put two more identical windings into the stator.
Each of the three windings could be kept electrically separate, with
their own pairs of terminals. We would then have three separate single-
phase alternators in the same space as the original. Each of these would

also have a good distribution factor of 0.966. The three winding emfs will of course be mutually out of phase with each other by 120°, since each whole winding will occupy 120° of stator space. What we now have is the basis of a three-phase alternator.

The term three-phase alternator is in some ways slightly misleading. What we have, in effect, are three identical single-phase alternators contained in the one machine. The three stator windings are brought out to their own separate pairs of terminals on the stator casing. These stator windings are referred to as phase windings, or phases. They are identified by the colours red, yellow and blue. Thus we have the red, yellow and blue phases. The circuit representation for the stator winding of such a machine is shown in Fig. 3.3. In this figure, the three phase windings are shown connected, each one to its own separate load. This arrangement is known as a three-phase, six-wire system. However, three-phase alternators are rarely connected in this way.

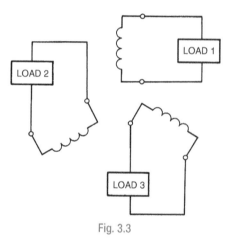

Fig. 3.3

Since the three generated voltages are sinewaves of the same frequency, mutually out-of-phase by 120°, then they may be represented both on a waveform diagram using the same angular or time axis, and as phasors. The corresponding waveform and phasor diagrams are shown in Figs. 3.4 and 3.5 respectively. In either case,

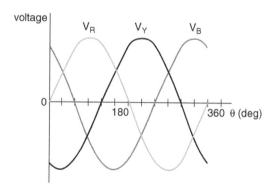

Fig. 3.4

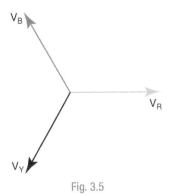

Fig. 3.5

we need to select a reference phasor. By convention, the reference is always taken to be the red phase voltage. The yellow phase lags the red by 120°, and the blue lags the red by 240° (or, if you prefer, leads the red by 120°). The windings are arranged so that when the rotor is driven in the chosen direction, the phase sequence is red, yellow, blue. If, for any reason, the rotor was driven in the opposite direction, then the phase sequence would be reversed, i.e. red, blue, yellow. We shall assume that the normal sequence of R, Y, B applies at all times.

It may be seen from the waveform diagram that at any point along the horizontal axis, the sum of the three voltages is zero. This fact becomes even more apparent if the phasor diagram is redrawn as in Fig. 3.6. In this diagram, the three phasors have been treated as any other vector quantity. The sum of the vectors may be determined by drawing them to scale, as in Fig. 3.6, and the resultant found by measuring the distance and angle from the beginning point of the first vector to the arrowhead of the last one. If, as in Fig. 3.6, the first and last vectors meet in a closed figure, the resultant must be zero.

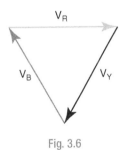

Fig. 3.6

3.2 Three-Phase, Four-Wire System

It is not necessary to have six wires from the three phase windings to the three loads, provided there is a common 'return' line. Each winding will have a 'start' (S) and a 'finish' (F) end. The common connection mentioned above is achieved by connecting the corresponding ends of the three phases together. For example, either the three 'F' ends or the three 'S' ends are commoned. This form of connection is shown in Fig. 3.7, and is known as a *star* or Y connection. With the resulting 4-wire system, the three loads also are connected in star configuration. The three outer wires are called the *lines*, and the common wire in the centre is called the *neutral*.

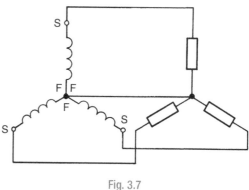

Fig. 3.7

If the three loads were identical in every way (same impedance and phase angle), then the currents flowing in the three lines would be identical. If the waveform and/or phasor diagrams for these currents were drawn, they would be identical in form to Figs. 3.4 and 3.5. These three currents meet at the star point of the load. The resultant current returning down the neutral wire would therefore be zero. The load in this case is known as a balanced load, and the neutral is not strictly necessary. However it is difficult, in practice, to ensure that each of the three loads are exactly balanced. For this reason the neutral is left in place. Also, since it has to carry only the relatively small 'out-of-balance' current, it is made half the cross-sectional area of the lines.

Let us now consider one of the advantages of this system compared with both a single-phase system, and the three-phase 6-wire system. Suppose that three identical loads are to be supplied with 200 A each. The two lines from a single-phase alternator would have to carry the total 600 A required. If a 3-phase, 6-wire system was used, then each line would have to carry only 200 A. Thus, the conductor csa would only need to be 1/3 that for the single-phase system, but of course, being six lines would entail using the same total amount of conductor material. If a 4-wire, 3-phase system is used there will be a saving on conductor costs in the ratio of 3.5:6 (the 0.5 being due to the neutral). If the power has to be sent over long transmission lines, such as the National Grid System, then the 3-phase, 4-wire system yields an enormous saving in cable costs. This is one of the reasons why the power generating companies use three-phase, star-connected generators to supply the grid system.

3.3 Relationship between Line and Phase Quantities in a Star-connected System

Consider Fig. 3.8, which represents the stator of a 3-phase alternator connected to a 3-phase balanced load. The voltage generated by each of the three phases is developed between the appropriate line and the neutral. These are called the phase voltages, and may be referred to in general terms as V_{ph}, or specifically as V_{RN}, V_{YN} and V_{BN} respectively. However, there will also be a difference of potential between any pair of lines. This is called a line voltage, which may be generally referred to as V_L, or specifically as V_{RY}, V_{YB} arid V_{BR} respectively.

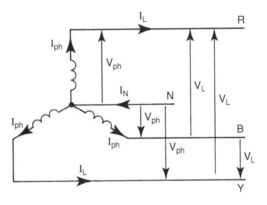

Fig. 3.8

A line voltage is the *phasor difference* between the appropriate pair of phase voltages. Thus, V_{RY} is the phasor difference between V_{RN} and V_{YN}. In terms of a phasor diagram, the simplest way to subtract one phasor from another is to reverse one of them, and then find the resulting phasor sum. This is, mathematically, the same process as saying that $a - b = a + (-b)$. The corresponding phasor diagram is shown in Fig. 3.9.

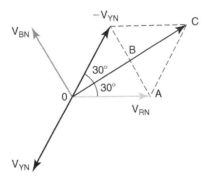

Fig. 3.9

Note: If V_{YN} is reversed, it is denoted either as $-V_{YN}$ or as V_{NY}. We shall use the first of these.

The phasor difference between V_{RN} and V_{YN} is simply the phasor sum of $V_{RN} + (-V_{YN})$. Geometrically this is obtained by completing the parallelogram as shown in Fig. 3.9. This parallelogram consists of two isosceles triangles, such as OCA. Another property of a parallelogram is that its diagonals bisect each other at right angles. Thus, triangle OCA consists of two equal right-angled triangles, OAB and ABC. This is illustrated in Fig. 3.10. Since triangle OAB is a $30°$, $60°$, $90°$ triangle, then the ratios of its sides AB:OA:OB will be $1:2:\sqrt{3}$ respectively.

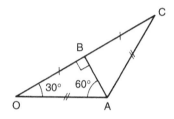

Fig. 3.10

Hence, $\dfrac{OB}{OA} = \dfrac{\sqrt{3}}{2}$

so $OB = \dfrac{\sqrt{3}.OA}{2}$

but $OC = 2 \times OB = \sqrt{3}.OA$

and since $OC = V_{RY}$, and $OA = V_{RN}$

then $V_{RY} = \sqrt{3}V_{RN}$

Using the same technique, it can be shown that:

$$V_{YB} = \sqrt{3}V_{YN} \text{ and } V_{BR} = \sqrt{3}V_{BN}$$

Thus, in star configuration, $V_L = \sqrt{3}V_{ph}$ (3.1)

The complete phasor diagram for the line and phase voltages for a star connection is shown in Fig. 3.11. Also, considering the circuit diagram of Fig. 3.8, the line and phase currents must be the same.

Hence, in star configuration, $I_L = I_{ph}$ (3.2)

We now have another advantage of a 3-phase system compared with single-phase. The star-connected system provides two alternative voltage outputs from a single machine. For this reason, the stators of all alternators used in electricity power stations are connected in star configuration. These machines normally generate a line voltage

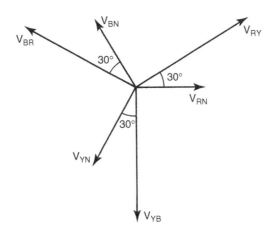

Fig. 3.11

of about 25 kV. By means of transformers, this voltage is stepped up to 400 kV for long distance transmission over the National Grid. For more localised distribution, transformers are used to step down the line voltage to 132 kV, 33 kV, 11 kV, and 415 V. The last three of these voltages are supplied to various industrial users. The phase voltage derived from the 415 V lines is 240 V, and is used to supply both commercial premises and households.

Worked Example 3.1

Q   A 415 V, 50 Hz, 3-phase supply is connected to a star-connected balanced load. Each phase of the load consists of a resistance of 25 Ω and inductance 0.1 H, connected in series. Calculate (a) phase voltage, (b) the line current drawn from the supply, and (c) the power dissipated.

A

Whenever a three-phase supply is specified, the voltage quoted is always the line voltage. Also, since we are dealing with a balanced load, then it is necessary only to calculate values for one phase of the load. The figures for the other two phases and lines will be identical to these.

$V_L = 415\,V; f = 50\,Hz; R_{ph} = 25\,\Omega; L_{ph} = 0.1\,H$

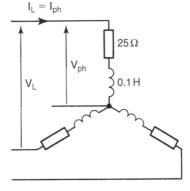

Fig. 3.12

(a) $V_{ph} = \dfrac{V_L}{\sqrt{3}} = \dfrac{415}{\sqrt{3}}$

so $V_{ph} = 240$ V **Ans**

(b)   Since it is possible to determine the impedance of a phase of the load, and we now know the phase voltage, then the phase current may be calculated:

$$X_L = 2\pi fL \text{ ohm} = 2\pi \times 50 \times 0.1$$

hence $X_L = 31.42\ \Omega$

$$Z_{ph} = \sqrt{R_{ph}^2 + X_L^2} \text{ ohm} = \sqrt{25^2 + 31.42^2}$$

$$Z_{ph} = 40.15\ \Omega$$

$$I_{ph} = \dfrac{V_{ph}}{Z_{ph}} \text{ amp} = \dfrac{240}{40.15}$$

so $I_{ph} = 5.98$ A

In a star-connected circuit, $I_L = I_{Ph}$
therefore $I_L = 5.98$ A **Ans**

The power in one phase, $P_{ph} = I_{ph}^2 R_{ph}$ watt

$$= 5.98^2 \times 25$$

$$P_{ph} = 893.29 \text{ W}$$

and since there are three phases, then the total power is:

$$P = 3 \times P_{ph} \text{ watt} = 3 \times 893.29$$

hence $P = 2.68$ kW **Ans**

3.4   Delta or Mesh Connection

If the start end of one winding is connected to the finish end of the next, and so on until all three windings are interconnected, the result is the delta or mesh connection. This connection is shown in Fig. 3.13. The delta connection is not reserved for machine windings only, since a 3-phase load may also be connected in this way.

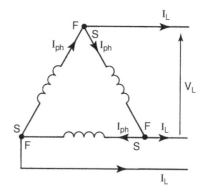

Fig. 3.13

3.5   Relationship between Line and Phase Quantities in a Delta-connected System

It is apparent from Fig. 3.13 that each pair of lines is connected across a phase winding. Thus, for the delta connection:

$$V_L = V_{ph} \qquad (3.3)$$

It is also apparent that the current along each line is the phasor difference of a pair of phase currents. The three phase currents are mutually displaced by 120°, and the phasor diagram for these is shown in Fig. 3.14. Using exactly the same geometrical technique as that for the phase and line voltages in the star connection, it can be shown that:

$$I_L = \sqrt{3}I_{ph} \qquad (3.4)$$

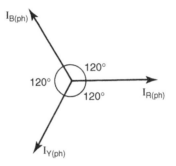

Fig. 3.14

The phasor diagram for the phase and line currents in delta connection is as in Fig. 3.15. Note that the provision of a neutral wire is not applicable with a delta connection. However, provided that the load

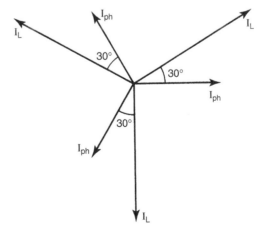

Fig. 3.15

is balanced, there is no requirement for one. Under balanced load conditions the three phase currents will be equal, as will be the three line currents. If the load is unbalanced, then these equalities do not exist, and each phase or line current would have to be calculated separately. This technique is beyond the scope of the syllabus you are now studying.

Worked Example 3.2

Q   A balanced load of phase impedance 120 Ω is connected in delta. When this load is connected to a 600 V, 50 Hz, 3-phase supply, determine (a) the phase current, and (b) the line current drawn.

A

$Z_{ph} = 120\ \Omega$; $V_L = 600\text{V}$; $f = 50\text{Hz}$

The circuit diagram is shown in Fig. 3.16.

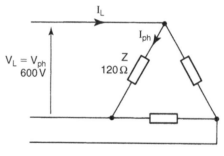

Fig. 3.16

(a)   In delta, $V_{ph} = V_L = 600$ V

$$I_{ph} = \frac{V_{ph}}{Z_{ph}}\ \text{amp} = \frac{600}{120}$$

so, $I_{ph} = 5$ A **Ans**

(b)   in delta, $I_L = \sqrt{3}I_{ph}\ \text{amp} = \sqrt{3} \times 5$

therefore $I_L = 8.66$ A **Ans**

3.6   Power Dissipation in Star and Delta-connected Loads

We have seen in Example 3.1 that the power in a 3-phase balanced load is obtained by multiplying the power in one phase by 3. In many practical situations, it is more convenient to work with line quantities.

$P_{ph} = V_{ph}I_{ph} \cos \phi$ watt

where $\cos \phi$ is the phase power factor.

total power, $P = 3 \times P_{ph}$

$$= 3 \times V_{ph}I_{ph} \cos \phi \text{ watt............[1]}$$

Considering a STAR-connected load,

$$V_{ph} = \frac{V_L}{\sqrt{3}}, \text{ and } I_{ph} = I_L$$

Substituting for V_{ph} and I_{ph} in eqn [1]:

$$P = 3 \times \frac{V_L}{\sqrt{3}} I_L \cos \phi$$

$$\text{therefore } P = \sqrt{3}V_L I_L \cos \phi \text{ watt} \qquad (3.5)$$

For a delta-connected load,

$$V_{ph} = V_L \text{ and } I_{ph} = \frac{I_{ph}}{\sqrt{3}}$$

and substituting these values into eqn [1] will yield the same result as shown in (3.5) above. Thus, the *equation* for determining power dissipation, in both star and delta-connected loads is exactly the same. However, the *value* of power dissipated by a given load when connected in star is not the same as when it is connected in delta. This is demonstrated in the following example.

Worked Example 3.3

Q   A balanced load of phase impedance 100 Ω and power factor 0.8 is connected (a) in star, and (b) in delta, to a 400 V, 3-phase supply. Calculate the power dissipation in each case.

A

$Z_{ph} = 100\,\Omega$; $\cos \phi = 0.8$; $V_L = 400\,V$

(a)   the circuit diagram is shown in Fig. 3.17

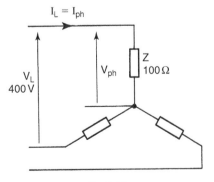

Fig. 3.17

$$V_{ph} = \frac{V_L}{\sqrt{3}} \text{ volt} = \frac{400}{\sqrt{3}}$$

so $V_{ph} = 231$ V

$$I_L = I_{ph} = \frac{V_{ph}}{Z_{ph}} \text{ amp} = \frac{231}{100}$$

and $I_L = 2.31$ A

$$P = \sqrt{3} \, V_L I_L \cos \phi \text{ watt} = \sqrt{3} \times 400 \times 2.31 \times 0.8$$

therefore $P = 1.28$ kW **Ans**

(b)   The circuit diagram is shown in Fig. 3.18.

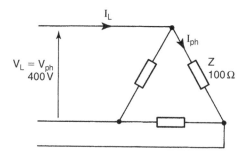

Fig. 3.18

$$V_{ph} = V_L = 400 \text{ V}$$

$$I_{ph} = \frac{V_{ph}}{Z_{ph}} \text{ amp} = \frac{400}{100}$$

so $I_{ph} = 4$A

but $I_L = \sqrt{3} I_{ph} = \sqrt{3} \times 4$

so $I_L = 6.93$ A

$$P = \sqrt{3} V_L I_L \cos \phi \text{ watt} = \sqrt{3} \times 400 \times 6.93 \times 0.8$$

therefore $P = 3.84$ kW **Ans**

Comparing the two answers for the power dissipation in the above example, it may be seen that:

> Power in a delta-connected load is *three times* that when it is connected in star configuration.   (3.6)

Worked Example 3.4

Q   A balanced star-connected load is fed from a 400 V, 50 Hz, three-phase supply. The resistance in each phase of the load is 10 Ω and the load draws a total power of 15 kW. Calculate (a) the line current drawn, (b) the load power factor, and (c) the load inductance.

A

$$V_L = 400\text{V}; f = 50\text{Hz}; R = 10\,\Omega; P = 15\,000\,\text{W}$$

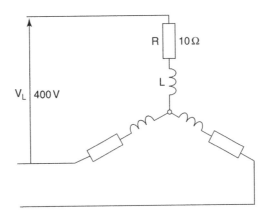

Fig. 3.19

(a) The power in one phase will be one third of the total power, so

$$P_{ph} = \frac{P}{3} \text{ watt} = \frac{15\,000}{3} = 5 \text{ kW}$$

but, $P_{ph} = I_{ph}^2 R$ watt

so, $I_L = I_{ph} = \sqrt{\frac{P_{ph}}{R}} = \sqrt{\frac{5000}{10}}$

$$I_L = 22.36 \text{ A } \textbf{Ans}$$

(b) $P = \sqrt{3}\, V_L I_L \cos \phi$ watt

so, $\cos \phi = \text{p.f.} = \frac{P}{\sqrt{3}\, V_L I_L} = \frac{15\,000}{\sqrt{3} \times 400 \times 22.36}$

hence, p.f. $= 0.968$ **Ans**

(c) $\phi = \cos^{-1} 0.968 = 14.47°$

$$\tan \phi = 0.2581 = \frac{X_{ph}}{R_{ph}}$$

so, $X_{ph} = 10 \times 0.2581 = 2.581$

$$L = \frac{X_L}{2\pi f} - \text{ohm} = \frac{2.581}{100\pi}$$

and $L = 8.22$ mH **Ans**

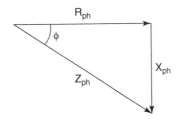

Fig. 3.20

Worked Example 3.5

Q   A balanced delta-connected load takes a phase current of 15 A at a power factor of 0.7 lagging when connected to a 115 V, 50 Hz, three-phase supply. Calculate (a) the power drawn from the supply, and (b) the resistance in each phase of the load.

A

$V_L = V_{ph} = 115\,V; f = 50\,Hz; I_{ph} = 15\,A; \cos\phi = 0.7$

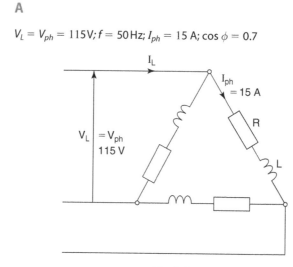

Fig. 3.21

(a) $I_L = \sqrt{3}I_{ph}\ \text{amp} = \sqrt{3} \times 15$

$I_L = 25.98\,A$

$P = \sqrt{3}\,V_L I_L \cos\phi\ \text{watt} = \sqrt{3} \times 115 \times 25.98 \times 0.7$

so $P = 3622.5\,W$ **Ans**

(b) $P_{ph} = \dfrac{P}{3} = \dfrac{3622.5}{3} = 1207.5\,W$

and $P_{ph} = I_{ph}^2 R\ \text{watt}$

so $R = \dfrac{P_{ph}}{I_{ph}^2}\ \text{ohm} = \dfrac{1207.5}{15^2}$

$R = 5.37\,\Omega$ **Ans**

3.7   Star/Delta Supplies and Loads

As explained in Section 3.3, the distribution of 3-phase supplies is normally at a much higher line voltage than that required for many users. Hence, 3-phase transformers are used to step the voltage down to the appropriate value. A three-phase transformer is basically three single-phase transformers interconnected. The three primary windings may be connected either in star or delta, as can the three secondary windings. Similarly, the load connected to the transformer secondary windings may be connected in either configuration. One important point to bear in mind is that the transformation ratio (voltage or turns ratio) refers to the ratio between the primary *phase* to the secondary *phase* winding. The method of solution of this type of problem is illustrated in the following worked example.

Worked Example 3.6

Q   Figure 3.22 shows a balanced, star-connected load of phase impedance 25 Ω and power factor 0.75, supplied from the delta-connected secondary of a 3-phase transformer. The turns ratio of the transformer is 20:1, and the star-connected primary is supplied at 11 kV. Determine (a) the voltages V_2, V_3 and V_4, (b) the currents I_1, I_2, and I_3, and (c) the power drawn from the supply.

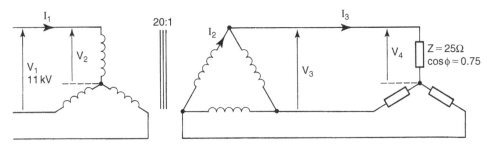

Fig. 3.22

A

$V_1 = 11\,000\,V$; $N_p/N_s = 20/1$; $Z_{ph} = 25\,\Omega$; $\cos\phi = 0.75$

(a) $$V_2 = \frac{V_1}{\sqrt{3}} = \frac{11\,000}{\sqrt{3}}$$

$V_2 = 6.351\,kV$ **Ans**

$$\frac{V_3}{V_2} = \frac{N_s}{N_p}$$

so $V_3 = \dfrac{N_s V_2}{N_p}$ volt $= \dfrac{6351}{20}$

hence $V_3 = 317.45\,V$ **Ans**

$$V_4 = \frac{V_3}{\sqrt{3}} = \frac{317.45}{\sqrt{3}}$$

and $V_4 = 183.3\,V$ **Ans**

(b)   In order to calculate the currents, we shall have to start with the load, and work 'back' through the circuit to the primary of the transformer:

$$I_3 = \frac{V_4}{Z_{ph}} \text{ amp} = \frac{183.3}{25}$$

$I_3 = 7.33\,A$ **Ans**

$$I_2 = \frac{I_3}{\sqrt{3}} = \frac{7.33}{\sqrt{3}}$$

$I_2 = 4.23\,A$ **Ans**

$$\frac{I_1}{I_2} = \frac{N_s}{N_p}$$

so $I_1 = \dfrac{N_s I_2}{N_p}$ amp $= \dfrac{4.23}{20}$

hence $I_1 = 0.212\,A$ **Ans**

(c) $$P = \sqrt{3}\, V_1 I_1 \cos\phi \text{ watt}$$
$$= \sqrt{3} \times 11 \times 10^3 \times 0.212 \times 0.75$$
therefore $P = 3.02$ kW **Ans**

Worked Example 3.7

Q   The star-connected stator of a three-phase, 50 Hz alternator supplies a balanced delta-connected load. Each phase of the load consists of a coil of resistance 15 Ω and inductance 36 mH, and the phase voltage generated by the alternator is 231 V. Calculate (a) the phase and line currents, (b) the load power factor, and (c) the power delivered to the load.

A

$f = 50\,\text{Hz}; R = 15\,\Omega; L = 36\,\text{mH}; V_{ph} = 231\,\text{V}$

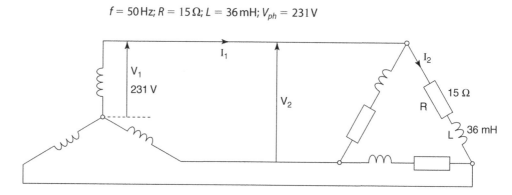

Fig. 3.23

(a) For the alternator: $V_1 = V_{ph} = 231$ V
$$V_2 = V_L = \sqrt{3}\, V_{ph}$$
$$V_2 = 400 \text{ V}$$
$$I_{ph} = I_L = I_1$$

For the load: $V_2 = V_L = V_{ph} = 400$ V
$$X_L = 2\pi f L \text{ ohm} = 100 \times \pi \times 36 \times 10^{-3}$$
$$X_L = 11.31\ \Omega$$
$$Z_{ph} = \sqrt{R^2 + X_L^2} \text{ ohm} = \sqrt{15^2 + 11.31^2}$$
$$Z_{ph} = 18.79\ \Omega$$
$$I_{ph} = I_2 = \frac{V_{ph}}{Z_{ph}} \text{ amp} = \frac{400}{18.79}$$
$$I_2 = 21.29 \text{ A } \textbf{Ans}$$
$$I_L = I_1 = \sqrt{3} I_2 = \sqrt{3} \times 21.29$$
$$I_1 = 36.88 \text{ A } \textbf{Ans}$$

(b) p.f. $= \cos\phi = \dfrac{R}{Z}$
$$= \frac{15}{18.79}$$
p.f. $= 0.8$ lagging **Ans**

(c) $$P = \sqrt{3}V_L I_L \cos\phi \text{ watt} = \sqrt{3} \times 400 \times 36.88 \times 0.8$$
$$P = 20.4 \text{ kW } \textbf{Ans}$$

Alternatively, $P = 3 \times P_{ph} = 3 \times I_p^2 R$ watt
$$= 3 \times 21.29^2 \times 15$$
$$P = 20.4 \text{ kW } \textbf{Ans}$$

3.8   Measurement of Three-phase Power

In an a.c. circuit the true power may only be measured directly by means of a wattmeter. The principle of operation of this instrument is described in *Fundamental Electrical and Electronic Principles,* Chapter 5. As a brief reminder, the instrument has a fixed coil through which the load current flows, and a moving voltage coil (or pressure coil) connected in parallel with the load. The deflection of the pointer, carried by the moving coil, automatically takes into account the phase angle (or power factor) of the load. Thus the wattmeter reading indicates the true power, $P = VI \cos\phi$ watt.

If a three-phase load is balanced, then it is necessary only to measure the power taken by one phase. The total power of the load is then obtained by multiplying this figure by three. This technique can be very simply applied to a balanced, star-connected system, where the star point and/or the neutral line are easily accessible. This is illustrated in Fig. 3.24.

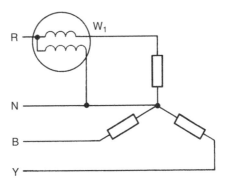

Fig. 3.24

In the situation where the star point is not accessible, then an artificial star point needs to be created. This is illustrated in Fig. 3.25, where the value of the two additional resistors is equal to the resistance of the wattmeter voltage coil.

In the case of an unbalanced star-connected load, one or other of the above procedures would have to be repeated for each phase in turn. The total power $P = P_1 + P_2 + P_3$, where P_1, P_2 and P_3 represent the three separate readings.

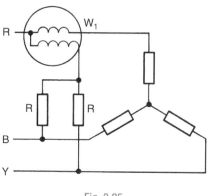

Fig. 3.25

For a delta-connected load, the procedure is not quite so simple. The reason is that the phase current is not the same as the line current. Thus, if possible, one of the phases must be disconnected to allow the connection of the wattmeter current coil. This is shown in Fig. 3.26. Again, if the load was unbalanced, this process would have to be repeated for each phase.

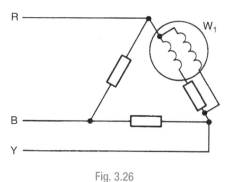

Fig. 3.26

3.9   The Two-Wattmeter Method

The measurement of three-phase power using the above methods can be very awkward and time-consuming. In practical circuits, the power is usually measured by using two wattmeters simultaneously, as shown in Fig. 3.27.

The advantages of this method are:

(a) Access to the star point is not required.
(b) The power dissipated in both balanced and unbalanced loads is obtained, without any modification to the connections.
(c) For balanced loads, the power factor may be determined.

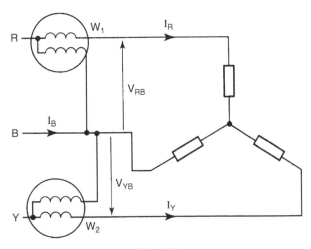

Fig. 3.27

Considering Fig. 3.27, the following statements apply:

Instantaneous power for $W_1, p_1 = v_{RB}i_R$ watt
and for $W_2, p_2 = v_{YB}i_Y$ watt
total instantaneous power $= p_1 + p_2$
$= v_{RB}i_R + v_{YB}i_Y$ ....[1]

Now, any line voltage is the phasor difference between the appropriate pair of phase voltages, hence:

$v_{RB} = v_{RN} - v_{BN}$ and $v_{YB} = v_{YN} - v_{BN}$

and substituting these into eqn [1] yields:

$$p_1 + p_2 = i_R(v_{RN} - v_{BN}) + i_Y(v_{YN} - v_{BN})$$
$$= v_{RN}i_R + v_{YN}i_Y - v_{BN}(i_R + i_Y)$$
but, $i_R + i_Y = -i_B$

i.e. the phasor sum of three equal currents is zero.

therefore $p_1 + p_2 = v_{RN}i_R + v_{YN}i_Y + v_{BN}i_B$

The instantaneous sum of the powers measured by the two wattmeters is equal to the sum of instantaneous power in the three phases.

Hence, total power,

$$P = P_1 + P_2 = P_R + P_Y + P_B \text{ watt} \qquad (3.7)$$

Consider now the phasor diagram for a resistive-inductive balanced load, with the two wattmeters connected as in Fig. 3.27. This phasor diagram appears as Fig. 3.28, below.

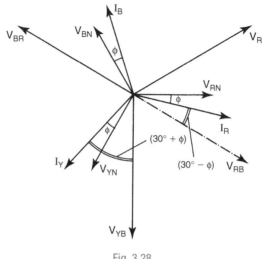

Fig. 3.28

The power indicated by W_1,

$$P_1 = V_L I_L \cos (30° - \phi) \tag{3.8}$$

and that for W_2,

$$P_2 = V_L I_L \cos (30° + \phi) \tag{3.9}$$

From these results, and using Fig. 3.29, the following points should be noted:

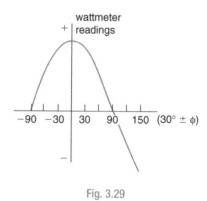

Fig. 3.29

1 If the load p.f. >0.5 (i.e. $\phi < 60°$); both meters will give a positive reading.

2 If the load p.f. $=0.5$ (i.e. $\phi = 60°$); W_1 indicates the total power, and W_2 indicates zero.

3 If the load p.f. <0.5 (i.e. $\phi > 60°$); W_2 attempts to indicate a negative reading. In this case, the connections to the voltage coil

of W_2 need to be reversed, and the resulting reading recorded as a negative value. Under these circumstances, the total load power will be $P = P_1 - P_2$.

4   The load power factor may be determined from the two wattmeter readings from the equation:

$$\phi = \tan^{-1}\sqrt{3}\left(\frac{P_2 - P_1}{P_2 + P_1}\right) \qquad (3.10)$$

hence, power factor, $\cos \phi$ can be determined.

Worked Example 3.8

Q   The power in a 3-phase balanced load was measured, using the two-wattmeter method. The recorded readings were 3.2 kW and 5 kW respectively. Determine the load power and power factor.

A

$$P_1 = 3.2 \text{ kW}; \quad P_2 = 5 \text{ kW}$$
$$P = P_1 + P_2 \text{ watt} = (3.2 + 5) \text{ kW}$$
therefore, $P = 8.2$ kW **Ans**

$$\phi = \tan^{-1}\sqrt{3}\left(\frac{P_2 - P_1}{P_2 + P_1}\right)$$

$$= \tan^{-1}\sqrt{3}\left(\frac{5 - 3.2}{5 + 3.2}\right)$$

hence, $\phi = 20.82°$
and p.f. $= \cos \phi = 0.935$ **Ans**

Worked Example 3.9

Q   A 3-phase balanced load takes a line current of 24 A at a lagging power factor of 0.42, when connected to a 415 V, 50 Hz supply. If the power dissipation is measured using the two-wattmeter method, determine the two wattmeter readings, and the value of power dissipated. Comment on the results.

A

$$I_L = 24 \text{ A}; \cos \phi = 0.42; V_L = 415\text{V}$$

$$\phi = \cos^{-1} 0.42 = 65.17°$$
$$P_1 = V_L I_L \cos (30° - \phi) \text{ watt}$$
$$= 415 \times 24 \times \cos (-35.17°)$$
therefore, $P_1 = 8.142$ kW **Ans**
$$P_2 = V_L I_L \cos (30° + \phi) \text{ watt}$$
$$= 415 \times 24 \times \cos (95.17°)$$
therefore, $P_2 = -896.7$ W **Ans**
$$P = P_1 + P_2 \text{ watt} = 8142 + (-896.7)$$
hence, $P = 7.244$ kW **Ans**

To obtain the negative reading on W_2, the connections to its voltage coil must have been reversed.

Worked Example 3.10

Q   A delta-connected load has a phase impedance of $100\,\Omega$ at a phase angle of $55°$, and is connected to a $415\,V$ three-phase supply. The total power consumed is measured using the two-wattmeter method. Determine the readings on the two meters and hence calculate the power consumed.

A

$Z_{ph} = 100\,\Omega; \phi = 55°; V_L = 415V = V_{ph}$

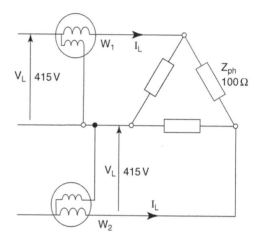

Fig. 3.30

$$I_{ph} = \frac{V_{ph}}{Z_{ph}}\ \text{amp} = \frac{415}{100} = 4.15\ \text{A}$$

$$I_L = \sqrt{3}I_{ph} = \sqrt{3} \times 4.15 = 7.19\ \text{A}$$

$$P_1 = V_L I_L\ \cos(30° - \phi)\ \text{watt} = 415 \times 7.19 \times \cos -25°$$

$$P_1 = 2.704\ \text{kW}\ \textbf{Ans}$$

$$P_2 = V_L I_L\ \cos(30° + \phi)\ \text{watt} = 415 \times 7.19 \times \cos 85°$$

$$P_2 = 260\ \text{W}\ \textbf{Ans}$$

$$P = P_1 + P_2\ \text{watt} = 2704 + 260$$

$$P = 2.964\ \text{kW}\ \textbf{Ans}$$

3.10   Neutral Current in an Unbalanced Three-phase Load

We have seen that the neutral current for a balanced load is zero. This is because the phasor sum of three equal currents, mutually displaced by $120°$, is zero. If the load is unbalanced, then the three line (and phase) currents will be unequal. In this case, the neutral has to carry the resulting out-of-balance current. This current is simply obtained by calculating the phasor sum of the line currents. The technique is basically the same as that used previously, by resolving the phasors into horizontal and vertical components, and applying Pythagoras' theorem. The only additional fact to bear in mind is that both horizontal and vertical components can have negative values. This is illustrated by the following example.

Worked Example 3.11

Q   An unbalanced, star-connected load is supplied from a 3-phase, 415V source. The three phase loads are purely resistive. These loads are 25 Ω, 30 Ω and 40 Ω, and are connected in the red, yellow and blue phases respectively. Determine the value of the neutral current, and its phase angle relative to the red phase current.

A

$V_L = 415$ V; $R_R = 25\,\Omega$; $R_Y = 30\,\Omega$; $R_B = 40\,\Omega$

The circuit diagram is shown in Fig. 3.31.

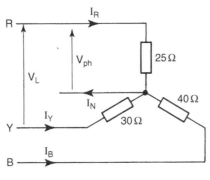

Fig. 3.31

$$V_{ph} = \frac{V_L}{\sqrt{3}} \text{ volt} = \frac{415}{\sqrt{3}} = 240 \text{ V}$$

$$I_R = \frac{V_{ph}}{R_R} \text{ amp; } I_Y = \frac{V_{ph}}{R_Y} \text{ amp; } I_B = \frac{V_{ph}}{R_B} \text{ amp}$$

$$= \frac{240}{25} \qquad = \frac{240}{30} \qquad = \frac{240}{40}$$

$$I_R = 9.6 \text{ A} \qquad I_Y = 8 \text{ A} \qquad I_B = 6 \text{ A}$$

The corresponding phasor diagrams are shown in Fig. 3.32.

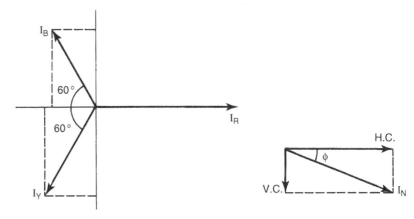

Fig. 3.32

Horizontal components, H.C. $= I_R - I_Y \cos 60° - I_B \cos 60°$

$$= 9.6 - (8 \times 0.5) = (6 \times 0.5)$$

so, H.C. $= 2.6$ A

Vertical components, V.C. $= I_B \sin 60° - I_Y \sin 60°$

$$= (6 \times 0.866) - (8 \times 0.866)$$

so V.C. $= -1.732$ A

The neutral current, $I_N = \sqrt{\text{V.C.}^2 + \text{H.C.}^2}$ amp

$$= \sqrt{-1.732^2 + 2.6^2}$$

hence, $I_N = 3.124$ A **Ans**

$$\phi = \tan^{-1}\frac{\text{V.C.}}{\text{H.C.}} = \tan^{-1}\frac{-1.732}{2.6}$$

hence, $\phi = -33.67°$ relative to I_R **Ans**

3.11   Advantages of Three-phase Systems

In previous sections, some of the advantages of three-phase systems, compared with single-phase systems, have been outlined. These advantages, together with others not yet described, are listed below. The advantages may be split into two distinct groups: those concerned with the generation and distribution of power, and those concerning a.c. motors. The last four of the advantages listed refer to the second group, and their significance will be appreciated when you have completed Chapter 5, which deals with a.c. machines.

1 The whole of the stator of a three-phase machine is utilised. A single phase alternator utilises only two thirds of the stator slots.
2 A three-phase alternator has a better distribution factor.
3 A three-phase, four-wire system provides considerable savings in cable costs, for the distribution of an equivalent amount of power.
4 A three-phase, four-wire system provides alternative voltages for industrial and domestic users.
5 For a given machine frame size, the power output from a three-phase machine is greater than that from a single-phase machine.
6 A three-phase supply produces a rotating magnetic field; whereas a single-phase supply produces only a pulsating field.
7 Three-phase motors are inherently self-starting; whereas single-phase motors are not.
8 The torque produced by a three-phase motor is smooth; whereas that produced by a single-phase motor is pulsating.

Summary of Equations

Phase and line quantities: In star, $I_L = I_{ph}$; $V_L = \sqrt{3}V_{ph}$

In delta, $V_L = V_{ph}$; $I_L = \sqrt{3}I_{ph}$

Power dissipation: $P = \sqrt{3}V_L I_L \cos\phi$ watt

$$P = 3 \times P_{ph} = 3 \times V_{ph}I_{ph}\cos\phi = 3 \times I_{ph}^2 R_{ph} \text{ watt}$$

and for the same load, $P(\text{delta}) = 3 \times P \text{ (star)}$

Power measurement: For two wattmeter method: $P = P_1 + P_2$ watt

$$\text{Load phase angle, } \phi = \tan^{-1}\sqrt{3}\left(\frac{P_2 - P_1}{P_2 + P_1}\right)$$

Assignment Questions

1   A three-phase load is connected in star to a 400 V, 50 Hz supply. Each phase of the load consists of a coil, having inductance 0.2 H and resistance 40 Ω. Calculate the line current.

2   If the load specified in Question 1 above is now connected in delta, determine the values for phase and line currents.

3   A star-connected alternator stator generates 300 V in each of its stator windings. (a) Sketch the waveform and phasor diagrams for the phase voltages, and (b) calculate the p.d. between any pair of lines.

4   A balanced three-phase, delta-connected load consists of the stator windings of an a.c. motor. Each winding has a resistance of 3.5 Ω and inductance 0.015 H. If this machine is connected to a 415 V, 50 Hz supply, calculate (a) the stator phase current, (b) the line current drawn from the supply, and (c) the total power dissipated.

5   Repeat the calculations for Question 4, when the stator windings are connected in star configuration.

6   Three identical coils, connected in star, take a total power of 1.8 kW at a power factor of 0.35, from a 415 V, 50 Hz supply. Determine the resistance and inductance of each coil.

7   Three inductors, each of resistance 12 Ω and inductance 0.02 H, are connected in delta to a 400 V, 50 Hz, three-phase supply. Calculate (a) the line current, (b) the power factor, and (c) the power consumed.

8   The star-connected secondary of a three-phase transformer supplies a delta-connected motor, which takes a power of 80 kW, at a lagging power factor of 0.85. If the line voltage is 600 V, calculate (a) the current in the transformer secondary windings, and (b) the current in the motor windings.

9   A star-connected load, each phase of which has an inductive reactance of 40 Ω and resistance of 25 Ω, is fed from the secondary of a three-phase, delta-connected transformer.

If the transformer phase voltage is 600 V, calculate (a) the p.d. across each phase of the load, (b) the load phase current, (c) the current in the transformer secondary windings, and (d) the power and power factor.

10 Three coils are connected in delta to a 415 V, 50 Hz, three-phase supply, and take a line current of 4.8 A at a lagging power factor of 0.9. Determine (a) the resistance and inductance of each coil, and (b) the power consumed.

11 The power taken by a three-phase motor was measured using the two-wattmeter method. The readings were 850 W and 260 W respectively. Determine (a) the power consumption and, (b) the power factor of the motor.

12 Two wattmeters, connected to measure the power in a three-phase system, supplying a balanced load, indicate 10.6 kW and −2.4 kW respectively. Calculate (a) the total power consumed, and (b) the load phase angle and power factor. State the significance of the negative reading recorded.

13 Using the two-wattmeter method, the power in a three-phase system was measured. The meter readings were 120 W and 60 W respectively. Calculate (a) the power, and (b) the power factor.

14 Each branch of a three-phase, star-connected load, consists of a coil of resistance 4 Ω and reactance 5 Ω. This load is connected to a 400 V, 50 Hz supply. The power consumed is measured using the two-wattmeter method. Sketch a circuit diagram showing the wattmeter connections, and calculate the reading indicated by each meter.

15 A three-phase, 415 V, 50 Hz supply is connected to an unbalanced, star-connected load, having a power factor of 0.8 lagging in each phase. The currents are 40 A in the red phase, 55 A in the yellow phase, and 62 A in the blue phase. Determine (a) the value of the neutral current, and (b) the total power dissipated.

Suggested Practical Assignments

Assignment 1

To determine the relationship between line and phase quantities in three-phase systems.

Apparatus:

Low voltage, 50 Hz, three-phase supply
3 × 1 kΩ rheostats
3 × ammeters
3 × voltmeters (preferably DMM)

Method:

1   Using a digital meter, adjust each rheostat to exactly the same value (1 kΩ).
2   Connect the rheostats, in star configuration, to the three-phase supply.
3   Measure the three line voltages and the corresponding phase voltages, and record your results in Table 1.
4   Measure the three line currents, and record these in Table 1.
5   Check that the neutral current is zero.
6   Carefully unbalance the load by altering the resistance value of one or two of the rheostats. ENSURE that you do not exceed the current ratings of the rheostats.
7   Measure and record the values of the three line currents and the neutral current, in Table 2.
8   Switch off the three-phase supply, and disconnect the circuit.
9   Carefully reset the three rheostats to their original settings, as in paragraph 1 of the Method.
10   Connect the rheostats, in delta configuration, to the three-phase supply, with an ammeter in each line.
11   Switch on the supply, and measure the line voltages and line currents. Record your values in Table 3.
12   Switch off the supply, and connect the three ammeters in the three phases of the load, i.e. an ammeter in series with each rheostat. This will involve opening each phase and inserting each ammeter.
13   Switch on and record the values of the three phase currents. Record values in Table 3.
14   From your tabulated readings, determine the relationship between line and phase quantities for both star and delta connections.
15   From your readings in Table 2, calculate the neutral current, and compare this result with the measured value.
16   Write an assignment report. Include all circuit and phasor diagrams, and calculations. State whether the line and phase relationships measured conform to those expected (allowing for experimental error).

Assignment 2

To measure the power in three-phase systems, using both single and two-wattmeter methods.

Apparatus:

Low voltage, 50 Hz, three-phase supply
3 × 1 kΩ rheostats
2 × wattmeters
1 × DMM

Method:

1. As for Assignment 1, carefully adjust the three rheostats to the same value (1 kΩ).
2. Connect the circuit shown in Fig. 3.33, and measure the power in the red and yellow phases.

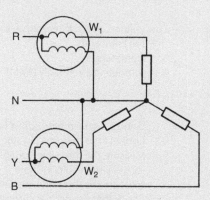

Fig. 3.33

3. Transfer one of the wattmeters to the blue phase, and measure that power.
4. Add the three wattmeter readings to give the total power dissipation. Check to see whether this is three times the individual phase power.
5. Switch off the power supply, and reconnect as in Fig. 3.34.

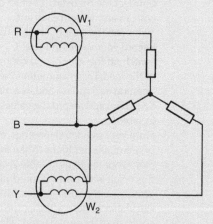

Fig. 3.34

6. Record the two wattmeter readings, and check whether their sum is equal to the total power recorded from paragraph 4 above.
7. Switch off the power supply and reconnect the circuit as in Fig. 3.35.
8. Record the two wattmeter readings.
9. Switch off the supply, and transfer one of the wattmeters to the third phase.
10. Record this reading. Add the three readings to give the total power, and check that this is three times the phase power.
11. Switch off the supply and connect the circuit of Fig. 3.36.
12. Record the two wattmeter readings, and check that their sum equals the total power obtained from paragraph 10 above.

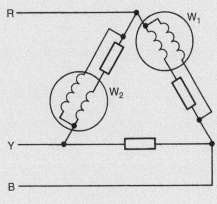

Fig. 3.35

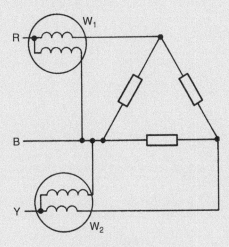

Fig. 3.36

Assignment 3

Show how the two-wattmeter method of power measurement can be accomplished by means of a single wattmeter, together with a suitable switching arrangement. You may either devise your own system, or discover an existing system through library research.

Chapter 4

Network Theorems

Learning Outcomes

These theorems provide an alternative to the application of Kirchhoff's laws to the solution of networks containing one or more sources of emf.

4.1 Superposition Theorem

The superposition principle allows us to break up a complex circuit into separate sections, and then apply simple Ohm's law techniques to each section in turn. The theorem is defined as follows:

> In any network consisting of resistors, the current flowing in any branch is the algebraic sum of the currents that would flow in that branch, if each emf source was considered separately, all others being replaced by resistors equal to their internal resistance.

In the above definition the term resistance has been used. This is because you are required to deal only with resistive circuits. However, the theorem applies equally to a.c. circuits, which may also include reactive elements. In this case, the word *impedance* should be substituted for resistance. This comment applies also to the other network theorems that follow shortly.

As with most theorems, it sounds most complicated. However, putting the principle into practice is relatively simple, and is best illustrated by means of an example.

Worked Example 4.1

Q   A circuit containing two sources of emf is shown in Fig. 4.1. Using the principle of Superposition, determine the current flowing in the 5 Ω resistor.

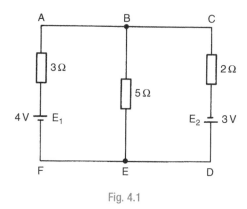

Fig. 4.1

A

The theorem allows us to split the original circuit into two sub-circuits (shown in Figs. 4.2 and 4.3), each of which contains only one battery, the other battery being replaced by its internal resistance. Note that it is not clear, from the diagram, whether the 3 Ω and 2 Ω resistors are the internal resistances of E_1 and E_2 respectively; or that E_1 and E_2 are ideal sources (having zero internal resistance), with the two resistors being external resistors in the circuit. In this case it makes no difference, so the two sub-circuits shown apply in both circumstances.

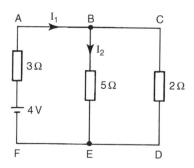

Fig. 4.2

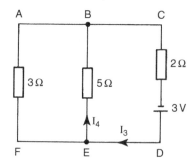

Fig. 4.3

Considering Fig. 4.2, currents I_1 and I_2 may be calculated as follows:

Total resistance of the circuit, R consists of the $3\,\Omega$ in series with the parallel combination of the $5\,\Omega$ and the $2\,\Omega$.

$$\text{Hence, } R = 3 + \frac{2 \times 5}{2 + 5} = 4.429\ \Omega$$

$$I_1 = \frac{E_1}{R}\ \text{amp} = \frac{4}{4.429}$$
$$\text{so, } I_1 = 0.903\ \text{A}$$

Using current divider technique, current I_2 is:

$$I_2 = \frac{2}{7} \times 0.903 = 0.258\ \text{A from B to E} \ldots\ldots\ldots[1]$$

Considering Fig. 4.3, we can apply the same techniques to evaluate currents I_3 and I_4:

$$R = 2 + \frac{5 \times 3}{5 + 3} = 3.875\ \Omega$$

$$I_3 = \frac{3}{3.875} = 0.774\ \text{A}$$

$$I_4 = \frac{3}{8} \times 0.774 = 0.29\ \text{A from E to B} \ldots\ldots\ldots[2]$$

The current through the $5\,\Omega$ resistor is the algebraic sum of the currents I_2 and I_4 as in [1] and [2] above. However, notice that these two currents are in opposite directions. The algebraic sum is therefore the *difference* between them. Hence, current through the $5\,\Omega = I_4 - I_2$ amp $= 0.032$ A from E to B **Ans**

4.2   Constant Voltage and Constant Current Sources and their Equivalence

Thus far, we have considered only voltage sources, in the form of a constant emf in series with its internal resistance. In some circuits it is more convenient to consider the source as a constant current generator, with its internal resistance in parallel with it. This is often the case when dealing with the equivalent circuit for a transistor.

Consider a constant voltage source of emf E volt and internal resistance r ohm, supplying a load resistor R, as shown in Fig. 4.4.

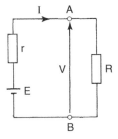

Fig. 4.4

$$\text{load current, } I = \frac{E}{r + R} \text{ amp}\ldots\ldots\ldots[1]$$

$$\text{and terminal p.d., } V = IR \text{ volt}\ldots\ldots\ldots[2]$$

Let the load R now be replaced by a short-circuit between terminals A and B, as shown in Fig. 4.5.

$$I_{sc} = \frac{E}{r} \text{ amp} \qquad\qquad [3]$$

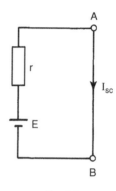

Fig. 4.5

Let us now replace the original voltage source by a constant current source, of value I_{sc} and internal resistance of r ohm in parallel with it. This is shown in Fig. 4.6. The linked circles form the circuit symbol for a current generator.

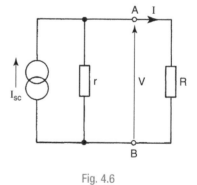

Fig. 4.6

Considering this figure, the current through the load, and the terminal p.d. may be determined by applying the current divider technique as follows:

$$I = \frac{r}{r + R} \times I_{sc} \text{ amp}$$

and substituting [3] for I_{sc}:

$$I = \frac{r}{r + R} \times \frac{E}{r}$$

so $I = \dfrac{E}{r + R}$ amp   [4]

and $V = IR$ volt   [5]

Since the expressions for the load current and terminal voltage in [1] and [2] are identical to those in [4] and [5] respectively, then it follows that the constant voltage generator of Fig. 4.4 is equivalent to the constant current generator of Fig. 4.6, and vice versa.

Worked Example 4.2

Q   (a) Convert the voltage source shown in Fig. 4.7 into its equivalent constant current source, and (b) convert the current source of Fig. 4.8 into its equivalent voltage source.

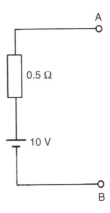

Fig. 4.7

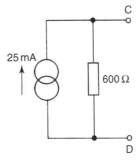

Fig. 4.8

A

(a) Placing a short-circuit between the terminals of the voltage generator will produce a current of:

$$I_{sc} = \frac{10}{0.5} \text{ amp} = 20 \text{ A}$$

so the constant current generator will produce a current of 20 A, with an internal resistance of 0.5 Ω in parallel, as shown in Fig. 4.9.

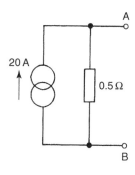

Fig. 4.9

(b) From Fig. 4.8, the open-circuit terminal voltage is:

$$V = 25 \times 10^{-3} \times 600 = 15 \text{ V}$$

so the constant voltage generator will have an emf of 15 V, with an internal resistance of 600 Ω in series, as shown in Fig. 4.10.

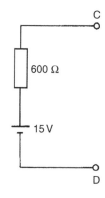

Fig. 4.10

4.3 Thévenin's Theorem

Thévenin's theorem states that any network containing sources and resistors can be replaced by a *single* voltage source of emf E_o and series internal resistance R_o, as shown in Fig. 4.11. The value of E_o is equal to the open-circuit terminal voltage of the network, measured between terminals A and B. The resistance R_o is the resistance of the network measured between terminals A and B, with all network sources replaced by their internal resistances.

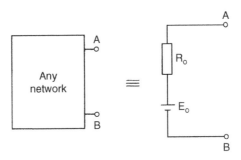

Fig. 4.11

Note: If any of the sources in the original network are shown as being ideal (zero internal resistance), these will be replaced by a short-circuit.

The application of Thévenin's theorem to the solution of a network is illustrated in the following example.

Worked Example 4.3

Q   For the network shown in Fig. 4.12, (a) determine the Thévenin equivalent generator for the circuit to the left of terminals A and B, (b) hence calculate the p.d. across and the current flowing through the 5 Ω load resistor, and (c) calculate the p.d. across and the current through the load if this resistance is changed to 3 Ω.

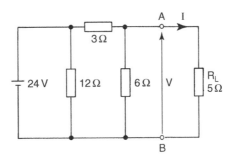

Fig. 4.12

A

(a)   With the 5 Ω resistor removed, terminals A and B will be open-circuit. The full 24 V will be developed across the 12 Ω resistor, so the 6 Ω and 3 Ω resistors form a simple potential divider.

Thus, $E_O = \dfrac{6}{9} \times 24 = 16$ V

Since the battery is shown as an ideal source, then it is replaced by a short-circuit. Thus, 'looking in' at the terminals, the circuit appears as in Fig. 4.13. As the 12 Ω resistor is now short-circuited, then the 6 Ω and the 3 Ω resistors are effectively connected in parallel.

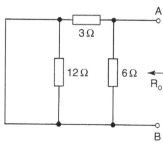

Fig. 4.13

Hence, $R_o = \dfrac{3 \times 6}{9} = 2\ \Omega$

The Thévenin equivalent generator therefore has an emf of 16V and internal resistance of 2 Ω. The complete equivalent circuit is shown in Fig. 4.14.

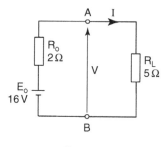

Fig. 4.14

(b)   From Fig. 4.14:

$$V = \frac{5}{7} \times 16 = 11.43 \text{ V } \textbf{Ans}$$

$$\text{and } I = \frac{V}{R_L} \text{ amp} = \frac{11.43}{5}$$

$$\text{so, } I = 2.29 \text{ A } \textbf{Ans}$$

(c) $$V = \frac{3}{5} \times 16 = 9.6 \text{ V } \textbf{Ans}$$

$$\text{and } I = \frac{9.6}{3} = 3.2 \text{ A } \textbf{Ans}$$

Note: The original circuit could have been solved using normal Ohm's law techniques, but the calculations would have been more extensive. The real advantage of deriving the Thévenin equivalent generator is that the load value may be changed to any value, as many times as is wished, but the subsequent calculations for *V* and *I* would be very quick and simple. Using normal Ohm's law techniques, *all* the circuit currents etc. would have to be recalculated each time.

Worked Example 4.4

Q   For the circuit of Fig. 4.15 (a) determine the Thévenin equivalent generator for the network to the left of terminals A and B, and hence calculate the current flowing through the 5 Ω load resistor, and (b) determine the load current if the load resistor is changed to 10 Ω.

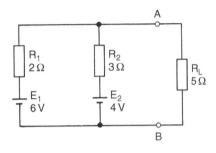

Fig. 4.15

A

(a)   Removing the load resistor, the circuit is as shown in Fig. 4.16, and the terminal p.d. (E_o) calculated as follows.

$$I = \frac{E_1 - E_2}{R_1 + R_2} \text{ amp} = \frac{6-4}{2+3}$$

therefore, $I = 0.4$ A

$$E_o = E_1 - IR_1 \quad \text{or} \quad E_2 + IR_2 \text{ volt}$$
$$= 6 - (0.4 \times 2) \quad \text{or} \quad 4 + (0.4 \times 3)$$

so, $E_o = 5.2$ V

From Fig. 4.17, where the two sources have been replaced by their internal resistances:

$$R_o = \frac{R_1 R_2}{R_1 + R_2} \text{ ohm} = \frac{6}{5} = 1.2 \ \Omega$$

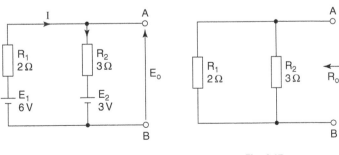

Fig. 4.16   Fig. 4.17

Hence, the Thévenin equivalent circuit will be as shown in Fig. 4.18, and from this circuit:

$$I = \frac{E_o}{R_o + R_L} \text{ amp} = \frac{5.2}{6.2}$$

so, $I = 0.839$ A **Ans**

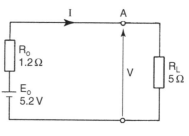

Fig. 4.18

(b) With R_L changed to $10\,\Omega$, and using the Thévenin circuit, the load current is simply obtained thus:

$$I = \frac{E_o}{R_o + R_L} \text{ amp} = \frac{5.2}{11.2} = 0.464 \text{ A } \textbf{Ans}$$

Note: This problem could have been solved using Kirchhoff's laws (or superposition theorem). However, in order to complete part (b), all of the loop equations would have to be derived again to provide a new set of simultaneous equations for solution. This clearly illustrates the advantage of using the Thévenin generator.

4.4   Norton's Theorem

This theorem complements Thévenin's in that the network is replaced by a constant current generator of I_{sc} amp, and parallel internal resistance R_o ohm. The current, I_{sc}, is the current that would flow between the terminals when they are short-circuited. The resistance R_o is defined in exactly the same way as it is for the Thévenin circuit. Thus R_o will be the same for both a Norton and a Thévenin equivalent circuit. The Norton constant current generator is illustrated in Fig. 4.19.

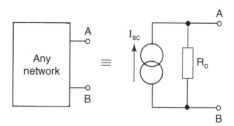

Fig. 4.19

The application of Norton's theorem is illustrated in the following example.

Worked Example 4.5

Q   Derive the Norton equivalent circuit for the network to the left of terminals A and B in Fig. 4.20, and hence determine the current through the $10\,\Omega$ load resistor.

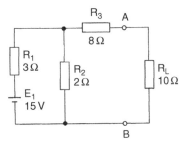

Fig. 4.20

A

With terminals A and B short-circuited, the circuit will be as in Fig. 4.21. Using this circuit, the short-circuit current I_{sc} is found as follows:

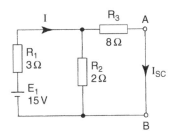

Fig. 4.21

$$\text{Total resistance, } R = R_1 + \frac{R_2\,R_3}{R_2 + R_3} \text{ ohm} = 3 + \frac{16}{10}$$

$$\text{so, } R = 4.6\ \Omega$$

$$I = \frac{E_1}{R} \text{ amp} = \frac{15}{4.6} = 3.261\ \text{A}$$

Using current division,

$$I_{sc} = \frac{R_2}{R_2 + R_3} \times I \text{ amp} = \frac{2}{10} \times 3.261$$

therefore, $I_{sc} = 0.652$ A

From Fig. 4.22, the resistance R_o is obtained thus:

$$R_o = R_3 + \frac{R_1\,R_2}{R_1 + R_2} \text{ ohm} = 8 + \frac{6}{5}$$

$$\text{so, } R_o = 9.2\ \Omega$$

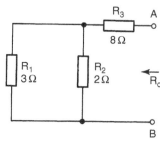

Fig. 4.22

The complete Norton equivalent circuit will be as shown in Fig. 4.23, and from this figure:

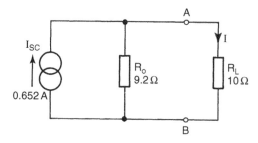

Fig. 4.23

$$I = \frac{R_o}{R_o + R_L} \times I_{sc} \text{ amp } \frac{9.2}{19.2} \times 0.652$$

hence, $I = 0.312$ A **Ans**

The advantage of the Norton equivalent circuit is the same as that for the Thévenin circuit; i.e. if the load resistance value is repeatedly changed, the new load current can be very simply and rapidly calculated.

Worked Example 4.6

Q   For the network shown in Fig. 4.24, determine (a) the Norton equivalent circuit, and (b) the Thévenin equivalent circuit.

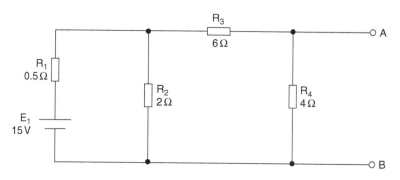

Fig. 4.24

A

(a) With terminals A and B short-circuited, R_4 is also short-circuited so the circuit becomes as shown in Fig. 4.25.

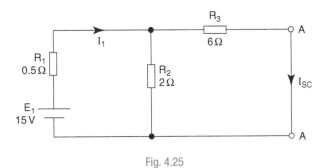

Fig. 4.25

Current from E_1 flows through R_1 and then through R_2 and R_3 in parallel, thus

total resistance, $R = R_1 + \dfrac{R_2 R_3}{R_2 + R_3}$ ohm $= 0.5 + \dfrac{6 \times 2}{6 + 2} = 2\ \Omega$

$$I_1 = \frac{E_1}{R} = \frac{15}{2} = 7.5\ \text{A}$$

Using current division, $I_{SC} = \dfrac{R_2}{R_2 + R_3} \times I_1$ amp

$$I_{SC} = \frac{2}{8} \times 7.5 = 1.875\ \text{A}$$

With terminals A and B open-circuited and with E_1 replaced by its internal resistance, R_1, the circuit is as shown in Fig. 4.26

$$R_{CD} = \frac{R_1 R_2}{R_1 + R_2} = \frac{0.5 \times 2}{2.5} = 0.4\ \Omega \qquad R_O = \frac{R_4 R_{ACD}}{R_4 + R_{ACD}} = \frac{4 \times 6.4}{10.64}$$

$$R_{ACD} = R_3 + R_{CD} = 6.4\ \Omega \qquad\qquad R_O = 2.462\ \Omega$$

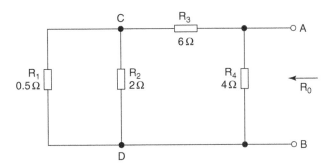

Fig. 4.26

The Norton equivalent circuit will therefore be as shown in Fig. 4.27 **Ans**

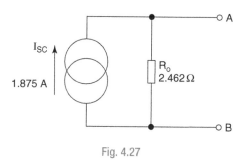

Fig. 4.27

(b) The circuit with terminals A and B open-circuited is shown in Fig. 4.28, and in this case current from E_1 will flow through R_1 and then split between R_2 and the series combination of R_3 and R_4.

$$\text{total resistance, } R = R_1 + \frac{R_2(R_3 + R_4)}{R_2 + R_3 + R_4} \text{ ohm}$$

$$R = 0.5 + \frac{2 \times 10}{12} = 2.1667 \ \Omega$$

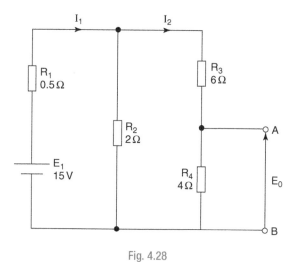

Fig. 4.28

$$\text{current drawn from the battery, } I_1 = \frac{E_1}{R} \text{ amp} = \frac{15}{2.1667}$$

$$I_1 = 6.923 \text{ A}$$

$$I_2 = \frac{R_2}{R_2 + R_3 + R_4} \times I_1 \text{ amp} = \frac{2}{12} \times 6.923$$

$$I_2 = 1.1538 \text{ A}$$

$$E_O = I_2 R_4 \text{ volt} = 1.1538 \times 4$$

$$E_O = 4.615 \text{ V}$$

The resistance of the Thévenin generator, R_o, is obtained in exactly the same manner as already done for the Norton equivalent, so the Thévenin equivalent circuit will be as shown in Fig. 4.29.

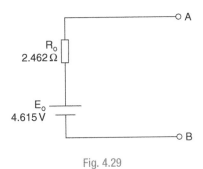

Fig. 4.29

Worked Example 4.7

Q   For the network shown in Fig. 4.30, (a) using Thévenin's theorem determine the current flowing through the $6\,\Omega$ resistor, and (b) verify your answer by using Norton's theorem.

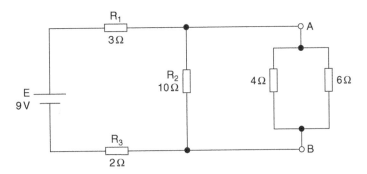

Fig. 4.30

A

(a) With terminals A and B open-circuited the circuit appears as in Fig. 4.31, and the total resistance R is:

$$R = R_1 + R_2 + R_3 \text{ ohm} = 3 + 2 + 10 = 15 \,\Omega$$

$$I = \frac{E}{R} \text{ amp} = \frac{9}{15} = 0.6 \text{ A}$$

$$E_O = V_{AB} = IR_2 \text{ volt} = 0.6 \times 10$$

$$E_O = 6 \text{ V}$$

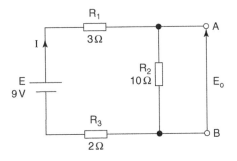

Fig. 4.31

Replacing E with its internal resistance $(0\,\Omega)$ the circuit is as in Fig. 4.32, and looking in at terminals A and B, it can be seen that R_2 is in parallel with the series combination of R_1 and R_3.

$$R_O = \frac{R_2(R_1 + R_3)}{R_2 + R_1 + R_3} \text{ ohm} = \frac{10 \times 5}{10 + 5}$$

$$R_O = 3.33\,\Omega$$

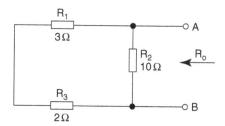

Fig. 4.32

The Thévenin equivalent circuit is shown in Fig. 4.33

$$R_{AB} = \frac{6 \times 4}{6 + 4} = 2.4\,\Omega$$

$$V_{AB} = \frac{R_{AB}}{R_{AB} + R_O} \times E_O = \frac{2.4}{2.4 + 3.33} \times 6$$

$$V_{AB} = 2.512\ V$$

$$I_{6\Omega} = \frac{V_{AB}}{6} \text{ amp} = \frac{2.512}{6}$$

$$I_{6\Omega} = 0.419\ A \textbf{ Ans}$$

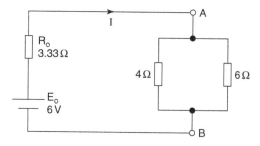

Fig. 4.33

(b) With terminals A and B short-circuited, the circuit will be as in Fig. 4.34, from which it may be seen that R_2 is also short-circuited. Thus the only opposition to the current flow will be the effect of R_1 and R_3 in series.

$$I_{SC} = \frac{E}{R_1 + R_3} \text{ amp} = \frac{9}{5} = 1.8\ A$$

The internal resistance of the Norton constant current generator, R_o, may be found in exactly the same manner as in part (a), so $R_o = 3.33\,\Omega$, and the Norton equivalent circuit will be as in Fig. 4.35.

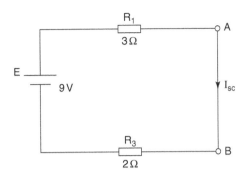

Fig. 4.34

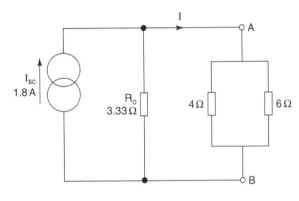

Fig. 4.35

Also from part (a) we know that $R_{AB} = 2.4\,\Omega$, so

$$I = \frac{R_O}{R_O + R_{AB}} \times I_{SC} \text{ amp} = \frac{3.33}{5.73} \times 1.8$$

$$I = 1.0465 \text{ A}$$

$$I_{6\Omega} = \frac{4}{10} \times 1.0465$$

$I_{6\Omega} = 0.419$ A **Ans** Which verifies the answer in part (a)

4.5   The Maximum Power Transfer Theorem

This theorem states that maximum power is transferred from a source to a load when the load resistance is of the same value as the internal resistance of the source.

Again, note that this theorem also applies to a.c. circuits; in which case the word *resistance* should be replaced by the word *impedance*. However, the generator may also possess internal reactance. In this case, for maximum power transfer, the load must also possess an equal value of reactance, but of the opposite type. In other words, if the source has inductive reactance, then the load must have an equivalent capacitive reactance, and vice versa.

Worked Example 4.8

Q   For the circuit shown in Fig. 4.20 (Example 4.5), determine the value of load resistor R_L that would result in maximum power dissipation in this load, and calculate the value of this power.

A

In order to solve this problem, the Norton (or Thévenin) equivalent circuit would be derived, as in Example 4.5. Thus, for this example, we know that the equivalent Norton generator has a current of 0.652 A, and internal resistance of 9.2 Ω. Hence the value of R_L that will result in maximum power transfer will be 9.2 Ω **Ans**.

The generator current I_{sc} will divide equally between R_o and R_L, hence:

$$\text{the load current, } I = \frac{0.652}{2} = 0.326 \text{ A}$$

$$P_L = I^2 R_L \text{watt} = 0.326^2 \times 9.2$$

$$\text{therefore } P_L = 0.978 \text{ W } \textbf{Ans}$$

In order to ensure that maximum power is transferred from a source to a load, the load has to be matched to the source. This effect can be achieved in a.c. circuits by means of an impedance matching transformer. Using this technique, the source is effectively connected to a load of impedance value equal to its own internal resistance. To illustrate the principle, consider a single-phase transformer, connected to a resistive load, as shown in Fig. 4.36.

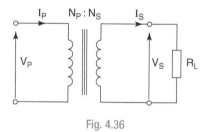

Fig. 4.36

The primary and secondary windings will be designed to have only a very small resistance (i.e. we may consider them to be perfect inductors). Thus the power developed in the secondary circuit must be due only to the secondary current flowing through the load resistor, R_L. However, the secondary of the transformer must derive its power from the primary circuit. The power in the primary circuit must therefore be due to the primary current flowing through some *effective* resistance. This effective resistance is known as 'the secondary resistance, referred to the primary'. In other words, the supply connected to the primary winding 'sees' an effective resistance connected between the primary terminals. The value of this referred resistance R'_p is obtained as follows.

$$R_L = \frac{V_s}{I_s} \text{ ohm; and } R'_p = \frac{V_p}{I_p} \text{ ohm}$$

$$\text{so, } \frac{R'_p}{R_L} = \frac{V_p}{I_p} \times \frac{I_s}{V_s} = \frac{V_p}{V_s} \times \frac{I_s}{I_p}$$

$$\text{but, } \frac{V_p}{V_s} = \frac{N_p}{N_s} \text{ and } \frac{I_s}{I_p} = \frac{N_p}{N_s}$$

$$\text{therefore, } \frac{R'_p}{R_L} = \left(\frac{N_p}{N_s}\right)^2$$

$$\text{hence, } R'_p = \left(\frac{N_p}{N_s}\right)^2 \times R_L$$

(4.1)

i.e. the resistance referred to the primary circuit is equal to the load resistance multiplied by the *square* of the turns ratio.

Worked Example 4.9

Q   A loudspeaker of impedance $8\,\Omega$ is fed from an amplifier of output impedance $1.8\,\text{k}\Omega$. Determine the turns ratio of a suitable matching transformer.

A

$$R_p = 1800\,\Omega; R_L = 8\,\Omega$$

$$R'_p = \left(\frac{N_p}{N_s}\right)^2 R_L \text{ ohm}$$

$$\text{so, } \left(\frac{N_p}{N_s}\right)^2 = \frac{R'_p}{R_L} = \frac{1800}{8}$$

$$\text{therefore, } \left(\frac{N_p}{N_s}\right)^2 = 225$$

$$\text{hence, } \frac{N_p}{N_s} = 15 \text{ i.e. turns ratio of 15:1 } \textbf{Ans}$$

Assignment Questions

1   Using the principle of superposition, determine the value of current flowing through the 6 Ω resistor of the circuit shown in Fig. 4.37.

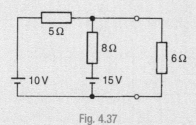

Fig. 4.37

2   By means of the superposition theorem, determine the value and direction of the current flow through the 10 Ω resistor in Fig. 4.38.

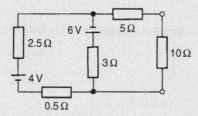

Fig. 4.38

3   Use the superposition theorem to calculate the p.d. and polarity between terminals A and B of the circuit shown in Fig. 4.39.

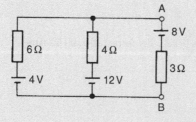

Fig. 4.39

4   Determine the Thévenin equivalent circuit for the network to the left of terminals A and B in Fig. 4.40. Hence calculate the current flowing through the 6 Ω resistor.

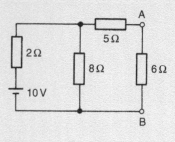

Fig. 4.40

5   Use Norton's theorem to calculate the current through the 6 Ω resistor of the circuit shown in Fig. 4.40.

6   Use Thévenin's theorem to solve Question 2 above.

7   For the network shown in Fig. 4.41, derive the Norton equivalent generator for the circuit to the left of terminals A and B, and hence calculate the p.d. across the 0.8 Ω resistor.

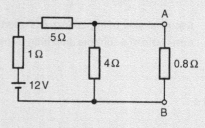

Fig. 4.41

8   For the network of Fig. 4.42, determine (a) the Thévenin equivalent circuit, (b) the Norton equivalent circuit, and (c) the value of resistor, connected between terminals X and Y, which will result in the transfer of maximum power from the source.

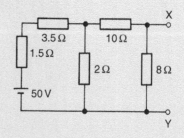

Fig. 4.42

9   Calculate the p.d. across and current through the 3.9 Ω resistor in Fig. 4.43.

Assignment Questions

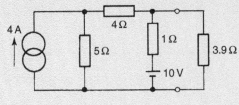

Fig. 4.43

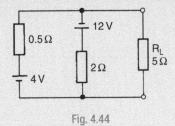

Fig. 4.44

10   For the circuit of Fig. 4.44, (a) using either
Thévenin's or Norton's theorem, determine
the p.d. across, and current through R_L, (b) the
value to which R_L must be changed to ensure
maximum power transfer into the load, and (c)
the value of this maximum power.

11   An a.c. source of internal impedance
600 Ω supplies a load of impedance 150 Ω.
Determine the turns ratio of the matching
transformer required for maximum power
transfer.

Chapter 5

A.C. Machines

Learning Outcomes

This chapter is concerned with the principles of operation of transformers, a.c. generators (alternators) and a.c. motors. On completion you should be able to:

1 Apply the emf equation for a transformer.
2 Understand the losses associated with a practical transformer, and hence explain the operation of the practical device.
3 Take measurements and calculate the efficiency of a transformer under varying load conditions.
4 Explain the principle of operation of single-phase and three-phase alternators, and carry out simple calculations concerning generated emf and efficiency.
5 Explain how a polyphase supply, when connected to a polyphase stator winding, produces a rotating magnetic field.
6 Explain the principles of polyphase and single-phase synchronous and induction motors, and the losses associated with these machines.
7 Explain the various methods for starting both polyphase and single-phase motors.
8 Explain the principles of simple stepping motors.

5.1 Transformer emf Equation

Figure 5.1 shows one complete cycle of the flux waveform in the core of a transformer. From this diagram the following statements can be made:

Total change of flux $= \Phi_m - (-\Phi_m) = 2\Phi_m$ weber

time taken for this change $= \dfrac{T}{2}$ second

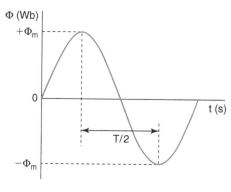

Fig. 5.1

therefore average rate of change of flux $= \dfrac{2\Phi_m}{T/2} = \dfrac{4\Phi_m}{T}$ Wb/s

in general, $e = -Nd\phi/dt$ volt, so for either winding:

average emf induced $= \dfrac{4N\Phi_m}{T}$ volt, and since $T = 1/f$, then

average emf induced $= 4\Phi_m Nf$ volt

and since we are dealing with sinewaves, having a form factor of 1.11, the r.m.s. emf induced is given by:

$$E = 4.44\Phi_m Nf \text{ volt} \qquad (5.1)$$

Notes:

1   The above equation applies to both primary and secondary induced emfs. For the primary, read E_p and N_p, and for the secondary, E_s and N_s.
2   The minus sign from the general equation has been omitted, but of course Lenz's law still applies. Thus E_p is in antiphase to the applied voltage V_p. Since E_p and E_s are induced by the same flux, then these two emfs are in phase with each other. Finally, since the secondary terminal voltage, V_s is produced by E_s, then V_s is antiphase to V_p. This confirms that the output of a transformer is phase inverted compared with its input.

Worked Example 5.1

Q   The primary winding of a single-phase transformer is connected to a 240 V, 50 Hz supply. The secondary winding has 1500 turns. If the peak value of the core flux is 1.965 mWb, determine (a) the number of turns on the primary winding, (b) the secondary induced emf, and (c) the core cross-sectional area if the flux density has a maximum value of 0.47 T.

A

$V_p = E_p = 240\,\text{V}; f = 50\,\text{Hz}; N_s = 1500; \Phi_m = 0.001965\,\text{Wb}; B_m = 0.47\,\text{T}$

(a)
$$E_p = 4.444\Phi_m N_p f \text{ volt}$$

therefore $N_p = \dfrac{E_p}{4.44\Phi_m f}$

$$= \dfrac{240}{4.44 \times 0.001965 \times 50}$$

hence $N_p = 550$ **Ans**

(b)
$$E_s = 4.44\Phi_m N_s f \text{ volt}$$

$$= 4.44 \times 0.001965 \times 1500 \times 50$$

hence $E_s = 654.3$ V **Ans**

(c)
$$B_m = \dfrac{\Phi_m}{A} \text{ tesla, so } A = \dfrac{\Phi_m}{B_m} \text{ metre}^2$$

therefore $A = \dfrac{0.001965}{0.47} = 41.81 \text{ cm}^2$ **Ans**

Worked Example 5.2

Q The 660-trurn primary winding of a single-phase transformer is connected to a 240V, 50 Hz supply, and it is required to provide an output of 40V. (a) Calculate the number of turns required on the secondary winding. (b) If the core cross-sectional area is 35 cm², calculate the maximum core flux density.

A

$N_p = 660; V_p = 240\text{V}; V_s = 40\text{V}; f = 50\text{Hz}; A = 35 \times 10^{-4} \text{ m}^2$

(a) $\dfrac{V_p}{V_s} = \dfrac{N_p}{N_s}$

$$N_s = \dfrac{V_s N_p}{V_p} = \dfrac{40 \times 660}{240}$$

$N_s = 110$ **Ans**

(b)
$$V_p = 4.44\Phi_m N_p f \text{ volt}$$

so, $\Phi_m = \dfrac{V_p}{4.44 N_p f} \text{ weber} = \dfrac{240}{4.44 \times 660 \times 50}$

$\Phi_m = 1.638$ mWb

$$B_m = \dfrac{\Phi_m}{A} \text{ tesla} = \dfrac{1.638 \times 10^{-3}}{35 \times 10^{-4}}$$

$B_m = 0.468$ T **Ans**

5.2   The Practical Transformer

The basic principles of operation for an ideal transformer are dealt with in *Fundamental Electrical and Electronic Principles*. These principles were briefly reiterated in Chapter 3 of this book, under the heading of maximum power transfer, where the principle of a matching transformer was described.

In the case of the matching transformer, it was stated that when a load is connected to the secondary winding, then a primary current is drawn. This must occur, since if the secondary supplies power to the load, then this power must be drawn from the a.c. supply connected to the primary. With the ideal machine, this input power will be the same value as that supplied to the secondary load. The phasor diagram for an ideal transformer, supplying a purely resistive load, is shown in Fig. 5.2. Note that, as with any phasor diagram, a reference phasor must be chosen. In the case of a transformer, the only quantity that is common to both the primary and secondary is the magnetic flux in the core. For this reason, the flux, Φ, is the reference phasor. Also note that a 2:1 step-down transformer has been chosen; hence V_p is twice V_s, and I_s is twice I_p. In addition, $E_p = V_p$ and $E_s = V_s$.

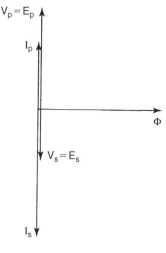

Fig. 5.2

If the load was now disconnected from the secondary, then I_s would be zero, and so too would I_p. However, the primary winding is still connected to the a.c. supply, so for a *practical* machine, some primary current must still be drawn. This is known as the primary no-load current, I_o. Thus, on no-load, the transformer is drawing power from the supply, even though it is not supplying any current to a load.

This no-load primary current is flowing through a highly inductive circuit (the primary winding), the resistance of which is designed to be as small as possible. Consequently, I_o will lag V_p by a large angle, ϕ_o. This current has two components, shown as I_{mag} and I_c in Fig. 5.3. I_{mag} is the magnetising component of the current, which is responsible for producing the flux in the core. The other component, I_c is in phase with V_p, and the product of these two quantities accounts for power

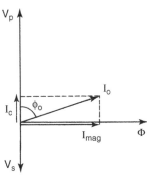

Fig. 5.3

loss in the transformer under these conditions. Since the resistance of the windings is small, and the no-load primary current is also relatively small, then the $I_o^2 R_p$ loss is insignificant. So what then is responsible for the power consumption? The answer to this question lies in the fact that the core flux is, of course, an alternating quantity. Consequently, the transformer core is being taken through continuous magnetisation cycles. This will result in both hysteresis and eddy current losses in the core. The effect of these losses is to cause unwanted heating of the 'iron' core, and are usually referred to collectively as the *iron losses*. The iron losses are kept to a minimum by making the core from thin laminations of a 'soft' magnetic material. The iron losses may be calculated from the equation shown below.

$$\text{Iron loss, } P_{Fe} = V_p I_o \cos \phi_o \text{ watt} \tag{5.2}$$

Worked Example 5.3

Q   A single-phase transformer has its primary winding connected to a 230 V, 50 Hz supply, and its secondary connected to a 250 Ω resistive load. Under this condition the current drawn from the supply is 4 A at unity power factor. Calculate (a) the secondary current supplied to the load, (b) the secondary voltage, and (c) the transformer turns ratio.

A

Without any information about losses we have to assume a perfect transformer, so the output power must equal the input power, and the primary and secondary phase angles will also be equal to each other.

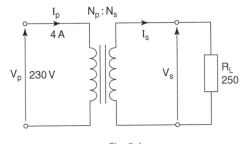

Fig. 5.4

(a) $P_o = P_i = V_p I_p \cos\phi$ watt

$ = 230 \times 4 \times 1 = 920$ W

also, $P_o = I_s^2 R_L$ watt

$$I_s = \sqrt{\frac{P_o}{R_L}} \text{ amp} = \sqrt{\frac{920}{250}}$$

$I_s = 1.918$ A **Ans**

(b) $P_o = V_s I_s \cos\phi$ watt

$$V_s = \frac{P_o}{I_s \cos\phi} \text{ volt} = \frac{920}{1.918}$$

$V_s = 479.7$ V **Ans**

(c) $\dfrac{N_s}{N_p} = \dfrac{V_s}{V_p} = \dfrac{479.7}{230}$

$\dfrac{N_s}{N_p} = \dfrac{2.086}{1} \approx \dfrac{2.1}{1}$

So turns ratio is 1:2.1 step up **Ans**

Let us now consider the practical transformer connected to a partly resistive, partly inductive load, as illustrated in Fig. 5.5. This will result in a secondary current, I_s, and a corresponding primary 'balancing' current, I_p'. However, there will still be the current I_o flowing in the

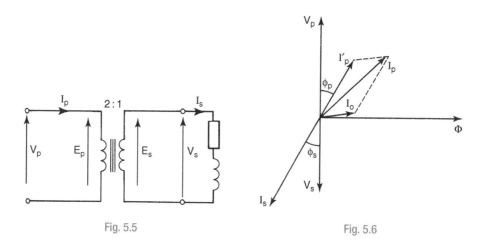

Fig. 5.5   Fig. 5.6

primary, since the core flux and core losses still exist. Hence the total primary current I_p must consist of the combination of these two currents. The appropriate phasor diagram is shown in Fig. 5.6. In this diagram, the *relative* size of I_o has been exaggerated simply for clarity. The practical consequence is that the primary phase angle, ϕ_p, is very nearly the same as that for the secondary, ϕ_s, and the current ratio is still taken to be the inverse of the turns ratio.

We have now seen that when the transformer is connected to a secondary load, the primary balancing current flows, so the total primary current increases in sympathy with the load current. The resistances of the two windings will now cause a more significant power loss, known as the *copper losses*. In addition, the impedance of each winding will introduce internal voltage drops, such that $E_p < V_p$ and $V_s < E_s$. This effect is illustrated in the phasor diagram of Fig. 5.7.

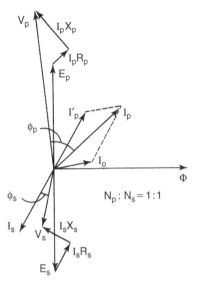

Fig. 5.7

Note: This phasor diagram is included here merely for completeness. You are not expected to be able to reproduce this diagram.

The copper, loss may be calculated from:

$$\text{Copper loss, } P_{Cu} = I_p^2 R_p + I_s^2 R_s \text{ watt} \qquad (5.3)$$

It should be borne in mind that despite the losses described, the transformer is the most efficient of all the electrical machines. Efficiency figures in the order of 95% to 99% are typical. The main reason for this high efficiency is the absence of any moving parts.

5.3   Measurement of Transformer Losses

The iron and copper losses may be determined by conducting two simple tests, known as the open-circuit and short-circuit tests. Once these losses are established, the efficiency can be calculated.

Open-Circuit Test

This test is used to measure the iron losses of the transformer. As the title implies, the secondary is left on open-circuit. The primary is connected to its normally rated supply voltage, as shown in Fig. 5.8. Under this no-load condition, the core flux will attain its normal value, but the primary current (I_o) will be about 5% of the full-load primary current (I_p). Under these circumstances the primary copper loss will be only 0.25% of the full-load value. Thus the wattmeter reading may be taken as the measurement of the core iron losses.

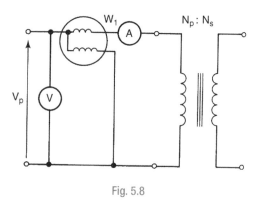

Fig. 5.8

Short-Circuit Test

This test is used to measure the copper losses when the transformer is supplying its rated full-load secondary current. Although this test is simple to apply, considerable care needs to be exercised in order to protect both the transformer and the supply from damage. The circuit arrangement is shown in Fig. 5.9, and the procedure is as follows.

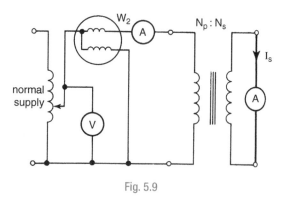

Fig. 5.9

A **variac** is simply a transformer having only one winding. The output is tapped off from a sliding contact. This device therefore acts as an a.c. potentiometer.

By means of the **variac** the primary voltage is very carefully increased, from zero, until the ammeter connected between the secondary terminals indicates the normally rated full-load output current. Under these conditions the primary voltage will be about 5% of its normal value, as will be the core flux. Since the iron loss is approximately proportional to the square of the flux, then this loss will be about 0.25% of its normal value. Hence the wattmeter reading may be taken as the transformer copper losses.

If the ammeter and voltmeter readings are noted when the above two tests are carried out, then the no-load and full-load power factors and phase angles may be determined.

$$\text{From the open-circuit tests: } \cos\phi_o = \frac{P_1}{V_p I_o}$$

$$\text{From the short-circuit test: } \cos\phi = \frac{P_2}{V_p I_p}$$

5.4   Transformer Efficiency

The efficiency of any device may be found by dividing its output power by the power supplied to it. The difference between the output and input must be the total losses of the machine. Thus, the efficiency may be expressed as:

$$\text{efficiency, } \eta = \frac{\text{output}}{\text{output} + \text{losses}}$$

The losses of the transformer are the iron and copper losses as measured by the open and short-circuit tests, and the full-load output is $V_s I_s \cos\phi$ watt. If we refer to these two forms of loss as P_{Fe} and P_{Cu} respectively, then the full-load efficiency is given by the expression:

$$\eta = \frac{V_s I_s \cos\phi}{V_s I_s \cos\phi + P_{Fe} + P_{cu}} \times 100\% \tag{5.4}$$

The above equation will give the efficiency under full-load conditions provided that $V_s I_s \cos\phi$ is the rated full-load output. When the transformer is supplying a smaller load its efficiency figure will not be the same value. The reason is that the copper losses, due to the resistance of the windings, will vary as the current varies. Thus the copper loss is a variable loss. On the other hand, the core losses are constant (fixed) regardless of the load supplied. The transformer will operate at its maximum efficiency when the variable loss equals the fixed loss. This is similar to the concept of maximum power transfer from a source to a load, i.e. when the load resistance (variable) equals the internal resistance (fixed) of the source. Hence, the equation for maximum efficiency is:

$$\eta_{max} = \frac{\text{output}}{\text{output} + 2 \times P_{Fe}} \times 100\% \qquad (5.5)$$

Worked Example 5.4

Q   A 20 kVA, single-phase transformer is tested on both open-circuit and short-circuit conditions, the losses thus measured being 120 W and 300 W respectively. Calculate the efficiency when supplying (a) full load at a power factor of 0.85, and (b) one quarter full load, at the same power factor.

A

From the open-circuit test, iron (fixed) losses, P_{Fe} = 120 W

From the short-circuit test, copper (variable) losses, P_{Cu} = 300 W

(a)   Full-load output, P_o = kVA cos ϕ watt

$$= 20 \times 10^3 \times 0.85$$

$$P_o = 17 \text{ kW}$$

$$\eta = \frac{P_o}{P_o + P_{Fe} + P_{Cu}} \times 100\%$$

$$= \frac{17\,000}{17\,000 + 120 + 300} \times 100\%$$

$$\eta = 97.6\% \text{ Ans}$$

(b)   Output on quarter half-load $= \dfrac{17\,000}{4} = 4250$ W

also, $P_{cu} \propto I_s^2$, and quarter load means the transformer is supplying a quarter of its normal full-load current, so on quarter load:

$$P_{Cu} \propto \left(\frac{I_s}{4}\right)^2 = \frac{I_s^2}{16}$$

so, $P_{Cu} = \dfrac{300}{16} = 18.75$ W

$$\eta = \frac{4250}{4250 + 120 + 18.75} \times 100\%$$

$$\eta = 96.8\% \text{ Ans}$$

5.5   Isolating Transformer

Since the input and output circuits of a transformer are electrically isolated from each other, a 1:1 turns ratio transformer may be used as an isolating transformer. This provides a degree of safety from the mains input supply, particularly when operating in hazardous conditions. Simple examples of this are the use of an isolating transformer in conjunction with a bathroom shaving socket, and the use of power tools on a building site.

5.6   The Auto Transformer

All the transformers so far considered have two separate windings, and are known as double-wound transformers. An auto transformer is one which has only one winding that is tapped to provide the output. A circuit diagram is shown in Fig. 5.10.

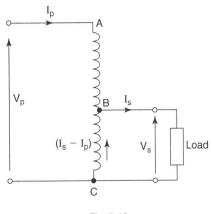

Fig. 5.10

The total winding between points A and C represents the primary winding of N_p turns. The tapped portion between B and C represents the secondary of N_s turns. Note that the current flowing through BC is the phasor difference between I_s and I_p. This is due to the fact that for any transformer the output is always antiphase to the input. Thus the current flowing through BC can be considerably less than that through AB.

Consequently the diameter of copper wire used to form section BC can be smaller than that used for AB. This in turn results in a reduction in weight and cost of materials required. Note that an auto transformer can only be used as a step-down voltage transformer.

The saving in copper used in an auto transformer compared to that used in an equivalent double-wound transformer can be calculated by the relationship

$$\text{(vol. of Cu used in auto trans)} = (1 - x)(\text{vol. of Cu used in dble-wnd transformer})$$

where $x = N_s/N_p$,

The saving in copper used is illustrated as follows. Suppose that the voltage ratio required is 4/3 step-down, which gives $x = 3/4 = 0.75$.

(vol. of Cu used in auto trans) $= (1 - x)$ (vol. of Cu used in dble-wnd transformer)
$= (1 - 0.75)$(vol. of Cu used in dble-wnd transformer)
$= 0.25$ (vol. of Cu used in dble-wnd transformer)

i.e. a saving in copper of 75%.

The auto transformer principle is used in the Variac. This is basically an a.c. potentiometer used to provide a variable voltage output ranging from the full input voltage to a small fraction of this value. However, since it is a variable device then it must be capable of carrying the full primary current, so the winding is of the same cross-section throughout and no saving in copper is possible.

5.7   Current Transformer

In order to measure large a.c. currents (say over 100 A) the direct use of a rectifier moving coil instrument is not possible due to the very large current that would be required to be diverted through the instrument shunt resistor. In this situation a current transformer may be employed, utilising the principle that $I_s = I_p(N_p/N_s)$ amp. Thus, if N_p is one or two turns only and N_s is a large number of turns, then the current passed through the meter will be only a small (known) fraction of that flowing in the load. This arrangement is illustrated in Fig. 5.11.

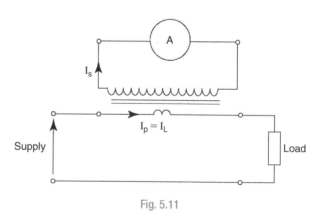

Fig. 5.11

When using a current transformer it is essential that the secondary is short-circuited before disconnecting the meter. If this is not done then a dangerously high voltage may be induced in the multi-turn secondary when the circuit is broken.

Worked Example 5.5

Q A single-phase transformer is rated at 10 kVA, 240 V/100 V. When tested the following results were obtained:

On open-circuit:

$V_p = 240$ V; $I_o = 2.6$ A; $P_{Fe} = 200$ W

On short-circuit:

$V_p = 18$ V; $I_p = 100$ A; $P_{Cu} = 250$ W

Using these results calculate (a) the no-load power factor, (b) the full-load efficiency, (c) the load at which maximum efficiency occurs, and (d) the value of maximum efficiency. Assume a load power factor of 0.8.

A

(a) $$P_{Fe} = V_p I_o \cos \phi_o \text{ watt}$$

so, $$\cos \phi_o = \frac{P_{Fe}}{V_p I_o} = \frac{200}{240 \times 2.6}$$

hence, $\cos \phi_o = 0.321$ **Ans**

(b) Since the transformer is rated at 10 kVA and the load power factor is 0.8, then the full-load power output is $10 \times 0.8 = 8$ kW.

full-load efficiency, $$\eta = \frac{\text{output}}{\text{output} + \text{losses}} \times 100\%$$

$$= \frac{8}{8 + 0.2 + 0.25} \times 100\%$$

therefore, $\eta = 94.67\%$ **Ans**

(c) The copper loss is directly proportional to the square of the current, and for maximum efficiency $P_{Cu} = P_{Fe} = 200$ W

Full-load current = 100 A;

so $100^2 \propto 250$ W .... [1]

and for η_{max}; $I^2 \propto 200$ W .... [2]

and dividing [2] by [1] we have:

$$I = \sqrt{\frac{200}{250}} \times 100^2 = 89.44 \text{ A}$$

$$P_o = 100 \times 89.44 \times 0.8$$

$$= 7155.4 \text{ W } \textbf{Ans}$$

(d) $$\eta_{max} = \frac{\text{output}}{\text{output} + 2 \times P_{Fe}} \times 100\%$$

$$= \frac{7155.4 \times 100}{7155.4 + 400}$$

hence, $\eta_{max} = 94.71\%$ **Ans**

Notice that there is very little difference between the full-load and maximum efficiency figures.

5.8   Alternators

The basic principles of operation of both single-phase and three-phase alternators has already been dealt with in Chapter 3, as an introduction to three-phase systems. It is suggested that you re-read the introductory section of that chapter to familiarise yourself with the construction of these machines.

5.9   Rotor Construction

The rotor of an alternator may be of one of two types: the salient pole type or the cylindrical type. The choice of construction depends on the speed of the prime mover which is used to drive the rotor.

Salient pole rotor

This form of construction is used when the driving speed is relatively low. Since the frequency of the emf generated, $f = np$ hertz, then for a given frequency, the lower the speed the greater the number of pole pairs required. For example, if the speed of rotation is 750 rev/min, then an 8-pole rotor would be required to generate at 50 Hz. This calculation is shown below.

$$f = 50 \text{ Hz}; n = \frac{750}{60} \text{ rev/s}$$

$$p = \frac{f}{n} = \frac{50 \times 60}{750} = 4$$

and number of poles $= 2p = 8$.

Fig. 5.12

In order to accommodate a large number of poles, the alternator needs to have a relatively large diameter. Also, since the power output of a machine is roughly proportional to its volume, this type of alternator would have a comparatively small axial length. Figure. 5.12 illustrates a portion of a salient pole rotor. The d.c. field winding would be contrawound on each alternate pole piece so as to provide alternate polarities (North–South–North etc.) at the tip of each. Each pole tip would also be well rounded to ensure a sinusoidal waveform.

Worked Example 5.6

Q   A 12-pole salient pole alternator is required to generate voltage at a frequency of 50 Hz. Calculate the speed of rotation in (a) rev/min, and (b) rad/s.

A

$p = 6; f = 50\,$Hz

(a) $f = np$ hertz

so, $n = \dfrac{f}{p}$ rev/second $= \dfrac{50}{6}$

$n = 8.33$ rev/s $= 8.33 \times 60$ rev/min

$n = 500$ rev/min **Ans**

(b) $\omega = 2\pi n$ rad/second $= 2\pi \times 8.33$

$\omega = 52.36$ rad/s **Ans**

Cylindrical rotor

The majority of alternators are driven by steam turbines, which are essentially high speed machines. In this situation a large number of poles is not necessary, and most large alternators of this type are two-pole machines, driven at 3000 rev/minute, resulting in a frequency of 50 Hz. A salient pole rotor would be inappropriate, since the centrifugal force acting on the poles would be very large. For example, the force on a 1 kg mass on the edge of a 1 m rotor, rotating at 3000 rev/min, would be about 50 kN. In the cylindrical rotor the field windings are contained in longitudinal slots machined into the outer periphery of the rotor core. Each slot contains a number of conductors which are securely fastened in by means of wedges. The basic construction is illustrated in Fig. 5.13. In addition to its great mechanical strength,

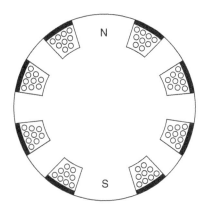

Fig. 5.13

the cylindrical rotor offers far less wind resistance (thus reducing the machine losses) and tends to produce a better flux pattern. The latter effect results in a more sinusoidal waveform than is generally available from a salient pole rotor.

5.10   Alternator emf Equation

Let z = number of stator conductors per phase
Φ = useful flux per pole, in Wb
p = number of *pairs* of poles
n = rotor speed, in rev/s

flux linking 1 conductor in 1 rev. = $2\Phi p$ weber
flux linking 1 conductor in 1 sec. = $2n\Phi p$, weber/second

and since a rate of change of flux of 1 Wb/s induces an emf of 1 volt, then:

$$\text{average emf generated in 1 conductor} = 2n\Phi p \text{ volt}$$
$$\text{average emf generated per phase} = 2p\Phi zn \text{ volt}$$

assuming a sinusoidal waveform (form factor of 1.11)

$$\text{then r.m.s. emf generated per phase} = 2.22p\Phi zn \text{ volt}$$

but frequency $f = np$ hertz, so the r.m.s. emf generated per phase is given by the equation:

$$E \text{ per phase} = 2.22\Phi zf \text{ volt} \tag{5.6}$$

Worked Example 5.7

Q   A 6-pole, single-phase alternator is driven at 1000 rev/min. There are 36 stator slots each containing 16 conductors. If the useful flux/pole is 0.04 Wb, calculate the frequency and r.m.s. value of the generated emf.

A

$p = 3; \Phi = 0.04\,\text{Wb}; n = 1000/60 \text{ rev/s}; z = 36 \times 16$

$$f = np \text{ hertz} = \frac{1000}{60} \times 3 = 50 \text{ Hz Ans}$$

$$E = 2.22\Phi zf \text{ volt} = 2.22 \times 0.04 \times 36 \times 16 \times 50$$

hence, $E = 2.557$ kV **Ans**

Worked Example 5.8

Q   A three-phase, 50 Hz alternator has a star-connected stator winding. There are 93 stator slots, each containing 4 conductors. If the useful flux/pole is 0.348 Wb, calculate the alternator line voltage.

A

$f = 50\,\text{Hz}; \Phi = 0.348\,\text{Wb}; z = 4 \times 93/3 = 124$

$$E_{ph} = 2.22\Phi zf \text{ volt} = 2.22 \times 0.348 \times 124 \times 50$$

$$= 4.79 \text{ kV}$$

$$E_L = \sqrt{3} \times E_{ph} = \sqrt{3} \times 4790$$

therefore, $E_L = 8.3$ kV **Ans**

Worked Example 5.9

Q   A 4-pole, single-phase alternator is required to generate an output of 230 V at a frequency of 50 Hz. The stator has a total of 78 slots, two thirds of which are filled with conductors. Determine (a) the speed of rotation, in rev/min, and (b) the number of conductors/slot required if the peak flux/pole is 10 mWb.

A

$$p = 2; \Phi_m = 0.01 \text{ Wb; slots used } = \frac{2}{3} \times 78 = 52; f = 50 \text{ Hz}$$

(a)   In order to generate a frequency of 50 Hz, the rotor speed will be

$$n = \frac{f}{p} \text{ rev/second} = \frac{50}{2}$$

$$n = 25 \text{ rev/s} = 25 \times 60 \text{ rev/min}$$

$$n = 1500 \text{ rev/min } \textbf{Ans}$$

(b) $E = 2.22\Phi_m zf$ volt

$$z = \frac{E}{2.22\Phi_m f}$$

$$= \frac{230}{2.22 \times 0.01 \times 50}$$

$$z = 207.2$$

However, we cannot have 0.2 of a conductor in a slot, so to ensure that 230 V are generated, there will need to be 208 conductors.

$$\text{Conductors/slot} = \frac{208}{52} = 4 \textbf{ Ans}$$

5.11 Alternator Losses

The input power to the alternator is in two parts; the mechanical driving power to the rotor, ωT watt (or $2\pi nT$), and the d.c. electrical power $V_r I_r$ watt which provides the excitation current for the field winding. The difference between the total input power and the electrical power output accounts for the losses in the machine. These losses consist of the iron, friction and windage losses, and the copper losses due to the resistance of the stator winding. The iron losses are due to hysteresis and eddy current losses in the stator core as the rotor field sweeps past it. There will also be mechanical losses due to friction in the bearings and slip-rings, together with the air friction (windage) of the rotor. These losses are best illustrated in the form of a power flow diagram as shown in Fig. 5.14.

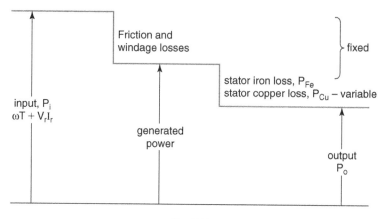

Fig. 5.14

The iron, friction and windage losses are sensibly constant, but the stator copper loss will naturally vary according to the amount of output current drawn by the load. Like the transformer, the alternator will operate at its maximum efficiency when the variable losses equal the fixed losses.

Worked Example 5.10

Q A 40 kVA, single-phase alternator supplies a full-load current of 100 A at a power factor of 0.85. Under this condition the iron, friction and windage loss is 1500 W, the stator copper loss is 4155 W, and the rotor field is supplied with 5 A d.c. at a p.d. of 100 V. Determine (a) the full-load efficiency, and (b) the input driving torque, if the speed is 3000 rev/min.

A

Output = 40 kVA; $\cos\varphi = 0.85$; $P_{Fe} = 1500\,W$; $n = 3000/60 = 50$ rev/s; $P_{cu} = 4155\,W$; $I_r = 5\,A$; $V_r = 100\,V$

(a)   output, P_o = kVA cos ϕ watt = 40 × 0.85

$$= 34 \text{ kW}$$

total losses = $P_{Fe} + P_{cu}$ watt

$$= 1500 + 4155 = 5655 \text{ W}$$

input, $P_i = P_o$ + losses = 34 000 + 5655

$$= 39\,655 \text{ W}$$

$$\eta = \frac{P_o}{P_i} \times 100\% = \frac{34\,000}{39\,655} \times 100$$

hence, $\eta = 85.74\%$ **Ans**

(b) $P_i = \omega T + V_r I_r$ watt

hence $T = \dfrac{P_i - V_r I_r}{\omega}$; where $\omega = 2\pi n$ rad/s

$$= \frac{39\,655 - 500}{100\pi}$$

therefore, $T = 124.6$ Nm **Ans**

Worked Example 5.11

Q   A 2-pole, single-phase alternator driven at 3000 rev/min requires a mechanical input power to its rotor of 30 kW when it is delivering its full-load output of 26.5 kW. Under this condition the fixed and variable losses are 950 W and 3 kW respectively, and the rotor field winding is supplied from a 110 V d.c. source. Given that the full-load efficiency is 87%, determine (a) the frequency generated, (b) the input driving torque on full-load, and (c) the rotor field current.

A

$p = 1; n = 3000$ rev/min; mech. input, $\omega T = 3 \times 10^4$ W; $P_o = 26.5 \times 10^3$ W

$P_{Fe} = 950$ W; $P_{Cu} = 3000$ W; $V_r = 110$ V; $\eta = 87\% = 0.87$

(a) $n = \dfrac{3000}{60}$ rev/s $= 50$ rev/s

$f = np$ hertz $= 50 \times 1$

$f = 50$ Hz **Ans**

(b) $\omega T = 3000$, where $\omega = 2\pi n$ rad/s

so, $\omega = 2\pi \times 50 = 100\pi$ rad/s

and, $T = \dfrac{3000}{100\pi}$

hence, $T = 9.55$ Nm **Ans**

(c) $\eta = \dfrac{P_o}{P_i}$

$P_i = \dfrac{P_o}{\eta}$ watt $= \dfrac{26.5 \times 10^3}{0.87}$

$P_i = 30.46$ kW

and from the power flow diagram shown in Fig. 5.14

$P_i = \omega T + V_r I_r$ watt

so, $V_r I_r = P_i - \omega T$ watt $= 30\,460 - 30\,000$

$V_r I_r = 460$ W

$I_r = \dfrac{460}{110}$

$I_r = 4.18$ A **Ans**

5.12   The Production of a Rotating Magnetic Field from a Polyphase Supply

In order to achieve the motor effect, there has to be an interaction between two magnetic fields. In the case of a.c. motors the stationary stator windings are required to produce a rotating flux pattern that will interact with the flux produced by the rotor winding. When a polyphase

supply is connected to an appropriate polyphase winding such a
rotating field results, as explained below.

Three-phase rotating field

Consider a three-phase supply connected to the stator windings of a
three-phase machine. For simplicity of explanation consider each phase
of the winding to be represented by a single-turn winding, as illustrated
in Fig. 5.15. Also shown in this diagram is the waveform of the supply
connected to the stator. The diagram of the stator has been repeated at
30° intervals in order to show the directions of the stator currents at
these instants of time. These current directions have been deduced by
using the following convention.

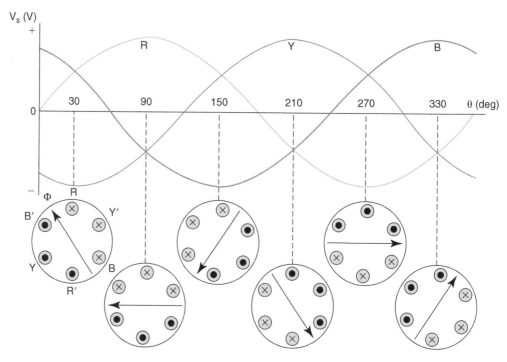

Fig. 5.15

When the relevant waveform is positive, the stator current will be
from R to R′, Y to Y′, or B to B′, as appropriate. Similarly, when
each waveform is negative, then the relevant current directions will
be reversed, i.e. R′ to R etc. It may be seen that the stator flux will
make one complete revolution during one complete cycle of the
supply. Thus the field will rotate at the same angular velocity as the
supply frequency. For example, if the supply frequency is 50 Hz, then
the stator field will rotate at 50 rev/s, or 100π rad/s. Another point to
note is that the *amplitude* of this rotating field is *constant*, with the
result that a three-phase motor will provide a smooth torque output.
Reference to these facts was made in Chapter 3, where the advantages
of three-phase systems compared with single-phase were stated.

Two-phase rotating field

A two-phase supply consists of two sinewaves that are 90° out of phase with each other. If such a supply is connected to two stator windings that are mutually displaced by 90°, then a rotating magnetic field is produced. This is illustrated in Fig. 5.16. A convention similar to that used for the three-phase system applies.

From Fig. 5.16 it may be seen that the rotating flux produced by the stator windings completes one revolution in one complete cycle of the supply. However, the amplitude of this flux is not constant, since at certain points only one winding is carrying current. For this reason the torque produced by a two-phase motor is not as smooth as that from a three-phase machine.

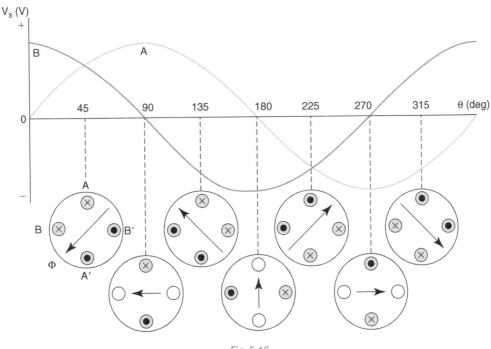

Fig. 5.16

Single-phase field

When a single-phase supply is connected to a single stator winding the flux produced will be pulsating, i.e. it will be sinusoidal. Thus the flux will change its polarity each half-cycle, and its amplitude will be continuously varying. However, such an alternating flux may be considered as consisting of two equal but contra-rotating flux phasors, each of amplitude equal to half the amplitude of the actual flux. The sequence of phasor diagrams shown in Fig. 5.17 illustrates the principle. It may be seen that, at any instant of time, the phasor sum of the two contra-rotating phasors (P_1 and P_2) is equal to the instantaneous value of the flux Φ. Due to the pulsating nature of the flux the torque produced by a single-phase motor is also pulsating.

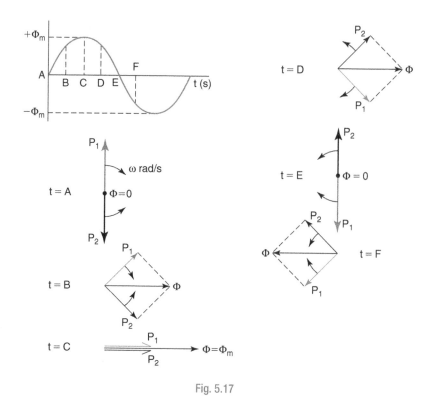

Fig. 5.17

In addition, a single-phase motor is not 'self-starting', as are the polyphase machines. This effect, and the means of overcoming the problem are covered later in this chapter.

5.13 Three-phase Induction Motor

The stator of this machine is identical in construction to that of a three-phase alternator. However, it is usual to bring the six ends of the three windings out to terminals mounted on the motor casing, enabling either star or delta connection, as required. The rotor construction may be of one of two types: a cage rotor or a wound rotor.

Cage rotor

This consists of aluminium bars contained in the slots of a cylindrical laminated steel core. The ends of the bars are connected together by aluminium shorting rings at each end of the rotor. In small to medium-sized machines these shorting rings are often cast with fins. These fins act as fan blades to circulate a cooling stream of air over the rotor. Figure 5.18 shows the cage construction, but the steel core has been omitted for the sake of clarity. There are no electrical connections to this form of rotor, and this type of machine is sometimes referred to as a brushless a.c. motor.

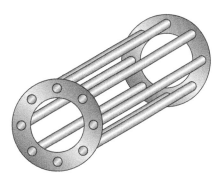

Fig. 5.18

Wound rotor

This also consists of a laminated steel core, but the slots around its periphery contain a three-phase winding that may be connected either in star or delta configuration. In some large machines the three ends of the rotor winding may be connected to slip-rings. This enables external connections to be made for starting purposes, which is explained later.

Operating principle

This is best understood by considering a single rotor bar on a cage rotor, with the stator field sweeping past it. This is illustrated in Fig. 5.19. The rotating stator field 'cuts' the bar, thus inducing an emf in it. Since the cage rotor forms a complete electrical circuit, then the induced emf causes current to flow, in the direction shown in Fig. 5.19. This direction of induced current may be confirmed by applying Fleming's right-hand rule. However, bear in mind that when this rule is applied it is assumed that the flux is stationary and the conductor is moving. Thus, in Fig. 5.19, the *relative* direction of motion of the conductor with respect to the flux is *anticlockwise*. The circulating rotor current produces its own magnetic field. The interaction between this and the stator field exerts a force on the rotor

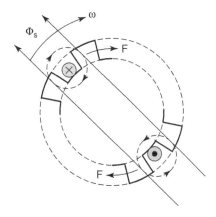

Fig. 5.19

which will cause it to rotate in the same direction as the stator field. In fact there will be a torque since the rotor bar diagonally opposite will have an opposite polarity current induced, as shown. As the flux pattern will tend to concentrate in the iron circuit, the torque is exerted on the rotor 'teeth', rather than on the rotor bars. The rotor must turn at some speed less than that of the stator field, otherwise there would be no 'cutting' action between this flux and the rotor bars; no induced emf and current; no interaction of fluxes; and hence no torque. The speed of rotation of the stator field is referred to as the synchronous speed. Thus the machine must always run at less than synchronous speed. However, when the machine is running light (no mechanical load connected) the rotor speed is very near to synchronous speed.

Slip

The difference between the rotor speed (n_r or ω_r) and the synchronous speed (n_s or ω_s) is known as the slip (s). This is usually expressed as a percentage, as follows:

$$\text{slip, } s = \frac{n_s - n_r}{n_s} \times 100\% \qquad (5.7)$$

It has been stated that on no-load n_r is very nearly equal to n_s, so the slip is small (maybe 0.1% or 0.2%). However, as the mechanical loading is increased (hence more torque required), the rotor speed falls. This is to be expected, since increased torque requires a stronger rotor flux. This effect can only be achieved by an increase of the 'cutting' action between the rotor and the stator field. This is further confirmed when considering the power flow diagram (Fig. 5.20). In order to

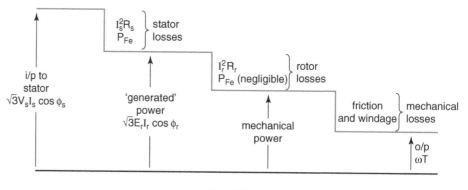

Fig. 5.20

provide more power output, the 'generated' power, $\sqrt{3}E_r I_r \cos \phi_r$ watt must be increased proportionately. The required increase in this power can only result from a larger induced emf, E_r. A typical torque/slip characteristic is shown in Fig. 5.21.

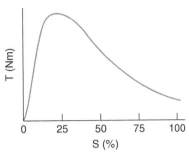

Fig. 5.21

The induction motor is the most commonly used a.c. motor, due to its simplicity of construction (cage rotor type) and its good speed/torque characteristics. Provided that it is not excessively loaded, the variation of speed is not too great, a full-load slip of about 6% being fairly typical.

Worked Example 5.12

Q   A three-phase cage rotor induction motor running at full-load draws a stator current of 60 A at a power factor of 0.8 lagging from a 415 V, 50 Hz supply. Under this condition the stator loss is 3.6 kW, the rotor loss is 1.3 kW, and the percentage slip is 4.5%. If the shaft output power is 29.4 kW, calculate (a) the shaft output torque, (b) the efficiency, and (c) the mechanical (friction and windage) loss.

A

$I_L = 60$ A; $\cos\phi = 0.8$; $V_L = 415$ V; $f = 50$ Hz; stator loss $= 3600$ W; rotor loss $= 1300$ W; $s = 4.5\% = 0.045$; $P_o = \omega_r T = 29\,400$ W

(a) $n_s = \dfrac{f}{p}$ rev/second, and since $p = 1$ for this machine

then, $n_s = 50$ rev/s

$$s = \frac{n_s - n_r}{n_s}$$

$$sn_s = n_s - n_r$$

$$n_r = n_s - sn_s = 50 - (0.045 \times 50)$$

$$n_r = 47.75 \text{ rev/s}$$

$$\omega_r = 2\pi n_r \text{ rad/s} = 2\pi \times 47.75 = 300 \text{ rad/s}$$

so, $T = \dfrac{29\,400}{300}$

$T = 98$ Nm **Ans**

(b) $P_i = \sqrt{3}\,V_L I_L \cos\phi$ watt $= \sqrt{3} \times 415 \times 60 \times 0.8$

$P_i = 34\,502$ W

$\eta = \dfrac{P_o}{P_i} \times 100\% = \dfrac{29\,400}{34\,502} \times 100\%$

$\eta = 85.2\%$ **Ans**

(c) From the power flow diagram of Fig. 5.20:

mech. loss = rotor mech. power − shaft output power ($\omega_r T$)

where rotor mech. power = P_i − (stator loss + rotor loss)

$$= 34\,502 - (3600 + 1300) = 29\,602 \text{ W}$$

so, mech. loss. = 29 602 − 29 400

mech. loss = 202 W **Ans**

5.14   Three-phase Synchronous Motor

The construction of this machine may be identical to that of a three-phase alternator. Indeed, if such an alternator had its stator connected to a suitable three-phase supply, and its rotor field winding was supplied with d.c. current, then the machine would operate as a synchronous motor.

Operating principle

To illustrate the basic principle consider a three-phase stator winding connected to a supply, but let the rotor be a small bar magnet mounted on a shaft. This is illustrated in Fig. 5.22. As the stator field sweeps past the magnet, the North pole of the stator field will attract the South pole of the magnet. Similarly, the diametrically opposite poles will also experience a force of attraction. Provided that the bar magnet has very little inertia it will accelerate rapidly and 'lock-on' to the rotating stator field. Thus our 'rotor' will continue to rotate at synchronous speed. This starting process will apply only when the machine has no mechanical load, and only if the rotor is very light, as just described. In a practical machine having the normal wound rotor, the inertia of the rotor inhibits this self-starting process. The reason is that it cannot accelerate quickly enough, so it is alternately attracted by and repelled

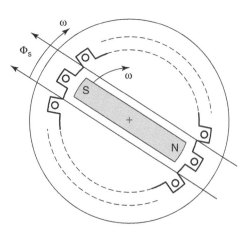

Fig. 5.22

by the stator field. The means of overcoming this problem will be described shortly.

Let us assume that on no-load the rotor and stator poles are perfectly aligned, as shown in Fig. 5.23(a). As the mechanical load is increased, this attempts to slow the rotor. The effect is to cause the rotor poles to move slightly out of perfect alignment as shown in Fig. 5.23(b). If the load is increased sufficiently, this misalignment reaches the point where the magnetic attraction is no longer capable of keeping the rotor locked to the stator field. This condition occurs at what is known as the 'pull-out' torque. The result is that the rotor slows down and stops. Thus the synchronous motor can run *only at synchronous speed*.

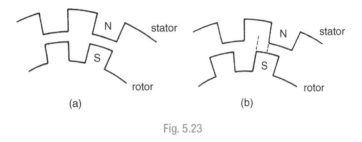

(a)   (b)

Fig. 5.23

Since this motor can operate only at one speed it is used in situations where a constant speed drive is required. The other useful feature of the synchronous motor is that, under appropriate loading conditions, it has a *leading* power factor. This is often used in industrial applications to provide an alternative means of power factor correction, instead of power factor correction capacitors. When used in this way the machine may be referred to as a *synchronous capacitor.* The added advantage is that the machine can also be utilised to perform other useful work at the same time. For example, it could be used to provide a constant speed drive for some process, such as driving ventilation fans. The power flow diagram is shown in Fig. 5.24.

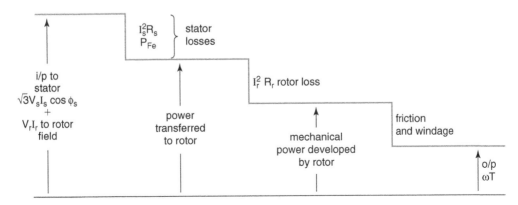

Fig. 5.24

5.15 Starting Methods for Three-phase Motors

For small motors (up to about 500 W) of both types, direct connection to the supply is possible. However, for large machines direct on-line starting is avoided. In the case of the synchronous motor self-starting is not possible due to rotor inertia, and in the case of the induction motor damage to both the machine and supply could result.

Synchronous motor starting

The simplest way to achieve this is to drive the rotor initially by means of an induction motor mounted on the same shaft. Under this condition the induction motor will be running on a very light load, so the slip will be small. Once the induction motor has reached this speed the supply to the synchronous motor stator is connected. The rotor of the synchronous motor will now be able to 'pull-in', 'lock-on' and continue running at synchronous speed. The supply to the induction motor can now be removed. With very large machines there may also be a clutch arrangement between the two motors. In this case, once the synchronous motor is running, the induction motor can also be mechanically disengaged.

Induction motor starting

Firstly let us consider the effect of connecting a large motor directly to the supply. At that instant the rotor would be stationary, so the slip would be 100%. The resulting 'cutting' action of the stator field would be very great. Since the resistance of the rotor will be very small (especially true for a cage rotor), then an excessively large emf and current would be induced in it. Now, an induction motor may be considered as a rotating transformer, whereby the primary winding is in the stator, and the rotor forms the secondary. Thus direct on-line starting would be akin to putting a short-circuit across the secondary of the transformer whilst it was connected to its normally rated supply. Under this condition the motor stator would draw an excessively large current from the supply. This could cause the stator windings to burn out and/or cause damage to the supply switchgear. Thus some means of reducing the current drawn at start-up is required. There are several methods used, two of which will now be described.

Star-Delta starter

In Chapter 3 we found that a star-connected load dissipates only one third of the power compared to when it is connected in delta. This fact is utilised in this form of starter. The stator of the induction motor, in this case, would be delta-connected under normal operating conditions. However, when starting the machine, these windings are connected in star configuration. Thus the input power and hence current drawn from the supply is limited. The six ends of the stator winding are brought out to terminals, which are then connected to a switching arrangement as shown in Fig. 5.25. Studying this diagram will show how the switchgear connects the winding firstly in star. Once the machine speed has built up, the stator winding is switched over to the delta connection, enabling the machine to develop its normal operating power. This method would have to be used for a cage rotor machine, but may also be used for the wound rotor type.

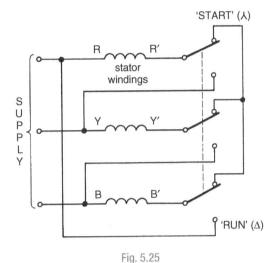

Fig. 5.25

Rotor resistance

For a wound rotor machine, an alternative method is to insert extra resistance into the rotor winding during the start-up period. Once the machine speed has stabilised this additional resistance is reduced to zero. To enable this process the ends of the rotor have to be connected to the external starter via slip-rings. The basic arrangement is illustrated in Fig. 5.26. For obvious reasons this method is not applicable to the cage rotor machine.

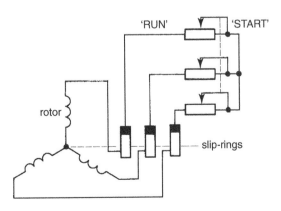

Fig. 5.26

5.16   Single-phase Motors

Although two-phase motors exist, they are used only in some specialist applications (such as servo motors) and will not be described here. However, the principle of a two-phase rotating field is utilised in some single-phase motors. The problems associated with a single-phase supply connected to a single stator winding were outlined in Section 5.12. The principle problem is that a single-phase motor is inherently not self-starting. Single-phase motors may be either synchronous or induction machines, and are generally small motors. They find their main usage in low-power applications and where a three-phase supply is not readily available. Examples are motors for mains operated electric clocks, office machinery, and some household appliances such as turntable drive motors for record players. Due to the simplicity of the cage rotor construction, the vast majority of single-phase motors are induction machines of this type.

Split-phase motors

These machines have two stator windings, mounted at 90° to each other. One of these, the 'start' winding (identified as S in Fig. 5.27) has a capacitor connected in series with it. Since the current through a capacitor leads the p.d. across it by 90°, then the current through this winding will be approximately 90° out of phase with respect to that flowing through the main winding, M. We therefore have the same effect as a two-phase supply connected to a two-phase stator winding. The result is that a rotating field is produced, and the machine is self-starting. A variation on this arrangement is to include a centrifugal switch, mounted on the rotor shaft, with its contact in series with the start winding. Once the machine has reached its operating speed, the centrifugal force causes the switch contact to open. The machine then continues to run on the main winding, 'following' the appropriate rotating flux phasor component, P_1 or P_2, described in Fig. 5.17. A typical application (without the centrifugal switch) is a central heating pump motor.

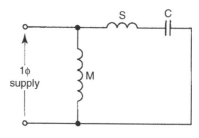

Fig. 5.27

Shaded pole motor

Each pole piece of this machine has a heavy copper ring embedded in one half, as illustrated in Fig. 5.28. Since a.c. is passed through the stator field windings, the flux is varying sinusoidally. The shading ring will therefore have a substantial eddy current induced into it. Being an induced current it obeys Lenz's law, and opposes the change of flux that induced it. The result is that the change of flux in one half of the pole piece is delayed compared with that in the other half. This is equivalent to the production of two flux components that are approximately 90° out of phase, which has the same effect as a two-phase field. The resultant flux pattern will therefore be rotating and this machine also is self-starting.

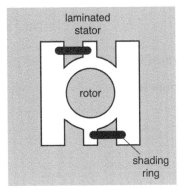

Fig. 5.28

Universal motor

This is actually a series d.c. motor, which can also operate from a single-phase a.c. supply. D.C. motors are fully described in the following chapter, but a brief explanation of this machine now follows. In a series motor the field winding is wound in series with the armature winding (the rotating part). Thus, as the current through the armature reverses each half cycle, so the polarity of the field is also reversed. The torque exerted on the armature therefore continues to act in one direction, and the armature is continuously rotated. This machine is the most commonly used single-phase motor. Typical applications are washing machines, vacuum cleaners, hair driers, hot-air hand driers, electric drills, etc. The main reason for the popularity of this motor is that it is simple to vary its speed. It is difficult and expensive to provide speed variation for true a.c. motors.

Summary of Equations

Transformer emf equation: $E = 4.44\Phi_m Nf$ volt

Practical transformer: Magnetising component of current, $I_{mag} = I_o \sin \phi_o$ amp

Power loss component of current, $I_c = I_o \cos \phi_o$ amp

Transformer losses: Iron loss, $P_{Fe} = V_p I_o \cos \phi_o$ watt
Copper loss, $P_{Cu} = I_p^2 R_p + I_s^2 R_s$ watt

Transformer efficiency: $\eta = \dfrac{V_s I_s \cos \phi}{V_s I_s \cos \phi + P_{Fe} + P_{Cu}} \times 100\%$

maximum efficiency occurs when $P_{Cu} = P_{Fe}$ (variable loss = fixed loss)

$$\eta_{max} = \frac{\text{output}}{\text{output} + 2 \times P_{Fe}} \times 100\%$$

Alternator generated frequency: $f = np$ hertz, where $p = pairs$ of poles

Alternator emf equation: $E_{ph} = 2.22\Phi_m zf$ volt

Assignment Questions

1. A transformer has a voltage ratio of 240 V/12 V. If there are 1500 turns on the primary winding, calculate the number of secondary turns.

2. A single-phase transformer is supplied at 100 V, and a secondary voltage of 440 V is required. If the secondary has 300 turns, determine the number of primary turns required.

3. A transformer supplies a load current of 9.5 A. If the primary to secondary turns ratio is 12:1, calculate the value of the primary current.

4. A step-up transformer has a turns ratio of 2:15. The primary is connected to a 240 V supply, and a 400 Ω resistive load is connected to the secondary. Calculate (a) the secondary voltage, (b) the primary current, and (c) the power drawn from the supply.

5. A 1:3 step-up transformer is connected to a 240 V, 50 Hz single-phase supply. If the number of secondary turns is 2700, calculate the peak value of the core flux.

6. A 50 Hz transformer has a core cross-sectional area of 480 cm^2 and a maximum flux density of 1.5 T. The number of turns on the primary and secondary windings are 75 and 250 respectively. Determine the primary and secondary induced emfs.

7. A 660 V/400 V, 25 kVA transformer yielded the following results when tested:

    On open-circuit:

    $V_p = 660$ V; $I_o = 2$ A; $P_1 = 240$ W

    On short-circuit:

    $V_p = 50$ V; $I_p = 103$ A; $P_2 = 500$ W

    Sketch the no-load phasor diagram and calculate (a) the no-load power factor, (b) the full-load efficiency, assuming a purely resistive load, and (c) the value of the total losses to ensure maximum efficiency.

8. A 10 kVA transformer has a primary voltage of 415 V, and produces a 100 V output. When connected to this supply it draws a no-load current of 1.2 A at a power factor of 0.3. On full-load, and supplying a purely resistive load, the copper losses are 350 W. Calculate (a) the full-load efficiency, (b) the copper loss that

    would result in maximum efficiency, and (c) the value of maximum efficiency. Sketch the no-load phasor diagram.

9. A 50 Hz alternator has 4 poles, and generates an emf of 800 V. If the useful flux per pole is 35 mWb, calculate (a) the number of conductors/phase, and (b) the speed at which it is driven.

10. A single-phase alternator generates an emf of 250 V at 50 Hz. There are 90 stator slots, two-thirds of which contain 4 conductors each. If it is driven at 1000 rev/min, calculate (a) the number of rotor poles, and (b) the useful flux/pole.

11. A three-phase, two pole, star-connected alternator is driven at 3000 rev/min. The stator has a total of 180 slots, each containing 5 conductors. If the useful flux/pole is 0.04 Wb, calculate (a) the generated emf per phase, (b) the line voltage, and (c) the output power when supplying a balanced load with a current of 15 A at 0.9 power factor.

12. With the aid of a power flow diagram, explain the losses of an alternator, identifying those losses that are fixed and those that are variable.

    A 15 kVA single-phase alternator supplies a load with 20 A, at a power factor of 0.78 lagging. The rotor is connected to a 50 V d.c. supply, from which it draws a current of 2 A. Under these conditions the fixed losses are 1.1 kW, and the machine has an efficiency of 80%. Determine (a) the mechanical power input, and (b) the stator copper losses.

13. (a) Explain the advantages of three-phase a.c. motors compared with single-phase motors.

    (b) With the aid of a simple sketch, explain how torque is developed on the rotor of an induction motor.

    (c) Explain the term 'slip', and state why this must occur in an induction motor.

14. Explain one method of starting a large synchronous motor, and one method for a large three-phase induction motor. Include in your answer the reasons why these starting methods are required.

Suggested Practical Assignments

Assignment 1

To determine the full-load efficiency for a single-phase power transformer.

Apparatus:

1 × single-phase power transformer
1 × wattmeter
1 × voltmeter
1 × variac
2 × ammeters (one of which must be capable of measuring the full-load secondary current)

Method:

1 Carry out the open and short-circuit test on the transformer. These techniques are outlined in Section 5.3.

    IMPORTANT NOTE: When conducting the short-circuit test it is **essential** that the variac is set to **zero volts** output, and this voltage is then increased **very carefully** until the ammeter connected across the secondary indicates the rated full-load secondary current.

2 Tabulate all meter readings from the two tests, and hence calculate the full-load efficiency of the transformer.

Assignment 2

To carry out a load test on a fractional horsepower three-phase induction motor.

Apparatus:

1 × 'low-voltage' 3-phase induction motor
1 × low-voltage 3-phase supply
2 × wattmeters
1 × tachometer
1 × brake test rig

Method:

1 Adjust the belt on the brake test rig so that it is just making contact with the brake drum attached to the motor shaft, and check that the spring balance on this rig indicates zero.

2 Connect the stator windings and the two wattmeters to the switched three-phase supply, as indicated in Fig. 5.29.

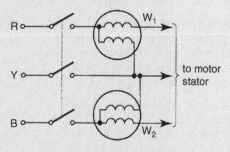

Fig. 5.29

3 Connect the supply and note the no-load speed of the motor by means of the tachometer.

4 Increase the mechanical loading on the motor, in steps, by increasing the brake belt tension. Note and tabulate the spring balance reading, the wattmeter readings and motor speed at each step.

5 Continue increasing the mechanical loading, as detailed in paragraph 4 above, until the machine's full rated power is indicated by the sum of the two wattmeter readings.

6 Calculate and tabulate the motor torque, slip, and output power for all loadings applied, using the following equations:

$$\text{motor torque, } T = Fr \text{ newton metre}$$

where F = spring balance reading, in newton
and r = radius of brake pulley, in metres

$$P_o = \omega T \text{ watt}$$

where $\omega = 2\pi n/60$ radian/second, and n = speed in rev/minute

7 Calculate the full-load efficiency of the motor.

8 Plot graphs of torque versus slip, and speed versus output power.

9 Complete a full assignment report.

Chapter 6

D.C. Machines

Learning Outcomes

This chapter covers the operating principles of d.c. generators and motors, their characteristics and applications. On completion you should be able to:

1 Appreciate the need for a commutator, the problems associated with the process of commutation and methods of overcoming these. Explain the rectifying action of a commutator.

2 Appreciate the problem of armature reaction, and methods used to reduce its affect. Describe the construction of d.c. machines.

3 Apply the emf equation and appreciate the effect on this equation of different types of armature winding.

4 Identify the different types of d.c. generator, and describe their characteristics. Carry out practical tests to compare the practical and theoretical characteristics.

5 Deduce the relationships between emf, torque, speed and current, and use these to deduce the characteristics of d.c. motors.

6 Explain the need for, and describe the operation of a d.c. motor starter.

7 List and explain the losses of d.c. machines. Carry out calculations involving these losses in order to predict the efficiency of d.c. machines.

8 Explain different methods of simple speed control for d.c. motors.

9 Explain the principles of simple stepper motors.

6.1   The Generation of D.C. Voltage

We have seen already that, if a single-loop coil is rotated between a pair of magnetic poles, then an *alternating* emf is induced into it. This is the principle of a simple form of alternator. Of course, this a.c. output could be converted to d.c. by employing a rectifier circuit. Indeed, that is exactly what is done with vehicle electrical systems. However, in order to have a truly d.c. machine, this rectification process needs to be automatically accomplished within the machine itself. This process is achieved by means of a commutator, the principle and action of which will now be described.

Consider a simple loop coil, the two ends of which are connected to a single 'split' slip-ring, as illustrated in Fig. 6.1. Each half of this slip-ring is insulated from the other half, and also from the shaft on which it is mounted. This arrangement forms a simple commutator,

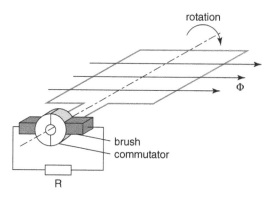

Fig. 6.1

where the connections to the external circuit are via a pair of carbon brushes. The rectifying action is demonstrated in the series of diagrams of Fig. 6.2. In these diagrams, one side of the coil and its associated commutator segment are identified by the shaded areas. For the sake of clarity, the physical connection of each end of the coil, to its associated commutator segment, is not shown. Figure 6.2(a) shows the instant when maximum emf is induced in the coil. The current directions have been determined by applying Fleming's right-hand rule. At this instant current will be fed out from the coil, through the external circuit from right to left, and back into the other side of the coil. As the coil continues to rotate from this position, the value of induced emf and current will decrease. Figure 6.2(b) shows the instant when the brushes

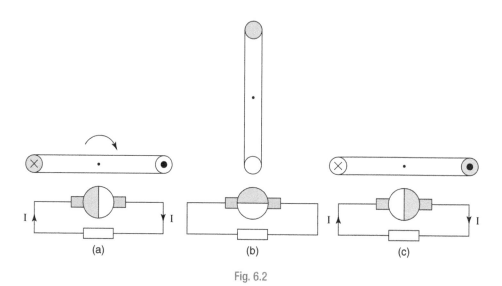

Fig. 6.2

short-circuit the two commutator segments. However, the induced emf is also zero at this instant, so no current flows through the external circuit. Further rotation of the coil results in an increasing emf, but of the opposite polarity to that induced before. Figure 6.2(c) shows the instant when the emf has reached its next maximum. Although the generated emf is now reversed, the current through the external circuit will be in the same direction as before. The load current will therefore be a series of half-sinewave pulses, of the same polarity. Thus the commutator is providing a d.c. output to the load, whereas the armature generated emf is alternating.

A single-turn coil will generate only a very small emf. An increased amplitude of the emf may be achieved by using a multi-turn coil. The resulting output voltage waveform is shown in Fig. 6.3. Although this emf is unidirectional, and may have a satisfactory amplitude, it is not a satisfactory d.c. waveform. The problem is that we have a concentrated winding. In a practical machine the armature has a number of multi-turn coils. These are distributed evenly in slots around the periphery

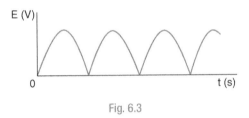

Fig. 6.3

of a laminated steel core. Each multi-turn coil has its own pair of slots, and the two ends are connected to its own pair of commutator segments. Figure 6.4 shows the armature construction (before the coils have been inserted). The riser is the section of the commutator to which the ends of the coils are soldered. Due to the distribution of the coils around the armature, their maximum induced emfs will occur one after the other, i.e. they will be out of phase with each other.

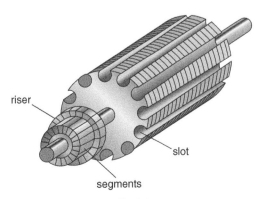

Fig. 6.4

Figure 6.5 illustrates this, but for simplicity, only three coils have been considered.

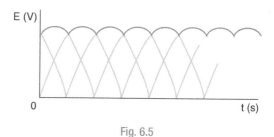

Fig. 6.5

Nevertheless, the effect on the resultant machine output voltage is apparent, and is shown by the thick line along the peaks of the waveform. With a large number of armature coils the ripple on the resultant waveform will be negligible, and a smooth d.c. output is produced.

6.2 The Commutation Process

Commutation is the process whereby the current in an armature coil is reduced to zero, reversed, and increased in the opposite direction. The problem is that this process needs to be carried out in the very short instant of time that the coil's armature segments are short-circuited by one of the carbon brushes. This problem is further aggravated by the inductance of the coils, since this will result in a delay in both growth and decay of current (refer to Chapter 8, dealing with d.c. transients). The commutation process is illustrated in Figs. 6.6 to 6.8 inclusive. Each of these diagrams represents a particular instant of time, as the commutator segments pass over a brush.

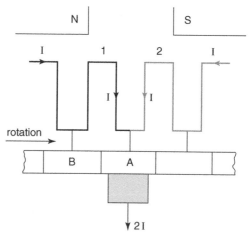

Fig. 6.6

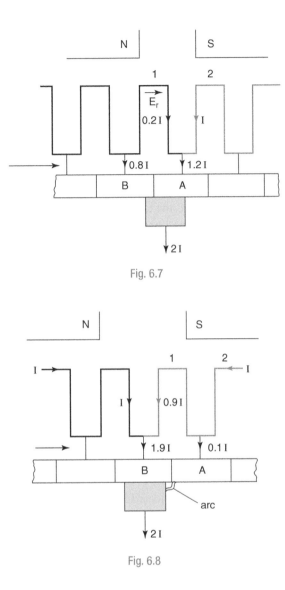

Fig. 6.7

Fig. 6.8

In Fig. 6.6, current from opposite halves of the armature winding arrive at the brush via coils 1 and 2 and segment A. Figure 6.7 shows the instant that segments A and B are short-circuited by the brush. At this instant coil 1 was carrying I amps, but as it is now short-circuited, the current needs to be zero. Thus the current in coil 1 is *trying* to reduce to zero instantaneously. This will induce a back emf, known as the reactance voltage E_r, into coil 1. This reactance voltage therefore opposes the decay of current. For illustrative purposes only, it has been assumed in Fig. 6.7 that the current has succeeded in decreasing to 20% of its original value at this time. In Fig. 6.8, segment A has just left contact with the brush. Ideally, under this condition, the full current of I amps should be flowing through coil 1, from the right-hand side of the armature. However, the reactance voltage, although decreased, will still be trying to maintain current flow in the opposite direction. For illustrative purposes it has been assumed that the current through coil 1 has succeeded in reversing to only 90% of its full value. Since Kirchhoff's current law must be maintained, the remaining 10% of the current from this side of the armature must reach the brush. The only way in which this can occur is for this proportion of the current to form an arc from segment A to the brush. This arcing will cause burning and pitting to both the commutator segments and the trailing edges of the brushes. This is obviously undesirable, and some means of reducing the sparking at the brushes is needed.

The key to the solution of the problem lies in the reactance voltage. If the short-circuited coil can have an emf induced in it that is equal and opposite to the reactance voltage, then there will be no opposition to the reversal of the coil current. One simple solution would be to advance the brushes, in the direction of rotation, so that the short-circuited coil comes under the influence of the next main pole ahead. Although this solution will reduce the sparking at the brushes, it also has another adverse effect. This will be described under the heading of armature reaction, in the next section of this chapter. However, for small motors and generators, the repositioning of the brushes is sometimes adopted.

6.3   Armature Reaction

Figure 6.9 represents an armature rotating between two poles. The direction of induced emfs will be as shown, but as the brushes are not connected to any external circuit, no armature current will flow. The magnetic neutral axis (MNA) is the axis along which zero emf is induced, and in this case it coincides with the geometric neutral axis (GNA).

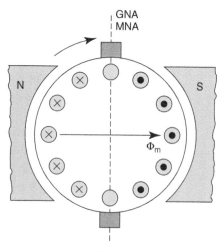

Fig. 6.9

If a load is now connected to the brushes, armature current will flow. This will result in a second magnetic field, the armature reaction flux Φ_A, as shown in Fig. 6.10. The interaction of this flux with the main field from the poles results in a resultant flux. This resultant is, in

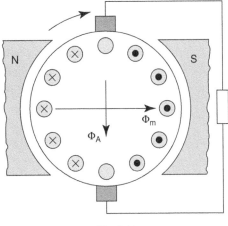

Fig. 6.10

effect, a distorted version of the main field, which results in a shift of the MNA through some angle θ degree in the direction of rotation. This effect is shown in Fig. 6.11.

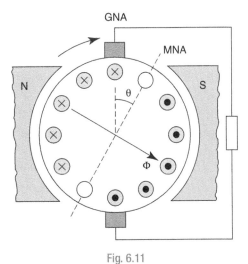

Fig. 6.11

For good commutation the brushes need to lie along the MNA, so that the brushes short-circuit armature coils only when zero emf is induced in them. Due to the armature reaction effect, brushes remaining on the GNA will be short-circuiting coils having emf induced in them. This will have the effect of reducing the overall emf available to the output, in addition to causing further arcing at the brushes. One solution to the problem would be to advance the brushes, through the angle θ, on to the new MNA. This solution would also provide the bonus of reducing the reactance voltage, thus reducing arcing due to commutation. However, there are two main problems with this solution. Firstly, the angle θ will vary as the load, and hence armature current, varies. Thus the brushes would have to be repositioned each time the load changed. This would not be very practical. Secondly, when the brushes are advanced, a band of armature conductors produce an mmf in direct opposition to the main field. This weakens the main field, and thereby reduces the value of generated emf. This band of demagnetising conductors is shown in Fig. 6.12.

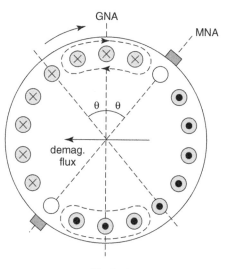

Fig. 6.12

For medium to large sized machines, the effects of armature reaction are reduced by the use of interpoles. These are small poles, interposed between the main poles. They are magnetised by the armature current that is passed through their windings. In the case of a generator, the polarity of each interpole will be that of the next main pole ahead, in the direction of rotation. This will have the effect of opposing the armature reaction flux, with the added benefit of reducing the reactance voltage due to commutation. The strength of the interpole flux required to cancel out armature reaction flux may not be that required to cancel the reactance voltage. In practice a compromise is achieved so that a considerable reduction in both is achieved. Note that the strength of the interpole flux will depend directly upon the load, and hence the value of armature current. Thus this flux will automatically compensate for variation of machine load. The effect of and positioning of interpoles is illustrated in Fig. 6.13.

A generator will operate as a motor, rotating in the same direction, if it is supplied with an armature current in the opposite direction to the generated current. The interpoles in such a machine will now have the

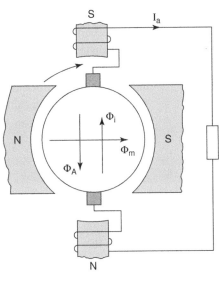

Fig. 6.13

polarity of the nearest main pole behind, with respect to the direction of rotation. This is the appropriate condition for a motor in order to overcome the armature reaction effect. Thus the inclusion of interpoles does not inhibit the generator/motor duality.

6.4   Construction of D.C. Machines

The various parts of a small d.c. machine are shown separately in Fig. 6.14, with the exception that neither the field nor armature windings

Fig. 6.14

have been included. The frame shell (bottom left) contains the pole pieces, around which the field winding would be wound. One end frame (top left) would simply contain a bearing for the armature shaft. The other end frame (bottom right) contains the brushgear assembly in addition to the other armature shaft bearing. The armature (top right) construction has already been described. The slots are skewed to provide a smooth starting and slow-speed torque for a motor.

6.5   Types of Armature Winding

There are two main forms of armature winding, known as wave and lap windings. The essential difference between them being the distance between the commutator segments to which the ends of the coils are connected.

Wave winding

Each coil of the winding spans one pole pitch. This means that one side of the coil is directly under one pole when the other side is directly under an opposite polarity pole. However, the two ends of the coil do not terminate at adjacent commutator segments. The arrangement is illustrated in Fig. 6.15. With the wave winding there are two parallel paths through the armature, regardless of the number of poles. If we

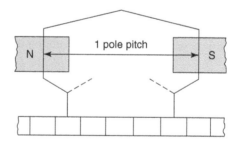

Fig. 6.15

consider each coil of the winding as an individual source of emf, then the machine may be represented by the equivalent circuit of Fig. 6.16. For the sake of explanation, let us assume that each source has an emf of 20 V, and produces a current of 5 A. It may be seen that the total machine emf is equal to the sum of the emfs of half of the coils connected in series. The total machine current is equal to twice the coil current. In the example shown, the machine emf would be 120 V, and the current would be 10 A.

Lap winding

With this winding each coil would again span one pole pitch, but the two ends would terminate at adjacent commutator segments. This

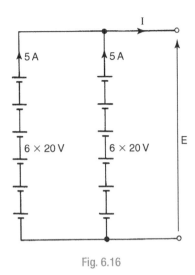

Fig. 6.16

is shown in Fig. 6.17. With this form of winding there are as many parallel paths through the armature as there are poles. If it is a two pole machine, then this will have the same effect as the wave winding. However, let us consider a four pole machine, and the armature equivalent circuit as shown in Fig. 6.18. Since there are four parallel paths, then the total machine emf will be 60 V, and the machine current will be 20 A.

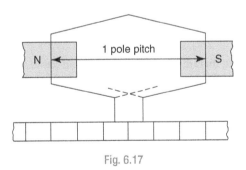

Fig. 6.17

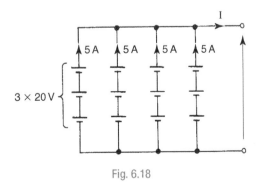

Fig. 6.18

Comparing these two types of winding, provided that the machine has four (or more) poles, then a wave wound armature provides 'high' voltage, 'low' current output; whereas the lap winding provides 'low' voltage, 'high' current output. Note that the terms 'low' and 'high' are used here only as relative terms when comparing the two winding types.

6.6   Generator emf Equation

Let z = total number of armature conductors

p = number of pole *pairs*

n = armature speed in rev/second

Φ = useful flux per pole

a = number of parallel paths through armature

Flux cut by one conductor in one rev. = $2p\Phi$ weber

flux cut by one conductor per second = $2p\Phi n$ weber

hence emf generated in one conductor = $2p\Phi n$ volt

since the machine has 'a' parallel armature paths, then the effective number of conductors is z/a, hence

$$\text{total generated machine emf, } E = \frac{2p\ \Phi zn}{a} \text{ volt} \quad (6.1)$$

Note: For a lap wound armature, $a = 2p$

and for a wave wound armature, $a = 2$

Worked Example 6.1

Q   A six-pole armature is wound with 480 conductors. The flux and speed are such that the emf generated in each conductor is 1.5 V, and the current flowing through each conductor is 100 A. Determine the total generated emf and power produced if the winding is (a) wave-wound, and (b) lap-wound.

A

$p = 3$; $z = 480$; $e = 1.5$ V; $i = 100$ A

(a) When wave-wound, number of parallel paths = 2, so there will be 240 conductors in each half.

Thus, total emf, $E = 240 \times e$ volt = 360 V **Ans**

and total current, $I = 2 \times i$ amp = 200 A **Ans**

and power output, $P = EI$ watt = 72 kW **Ans**

(b) When lap-wound, number of parallel paths = $2p = 6$, so there will be 480/6 = 80 conductors per path.

Thus, total emf, $E = 80 \times e$ volt = 120 V **Ans**

and total current, $I = 6 \times i$ amp = 600 A **Ans**

and power output, $P = EI$ watt = 72 kW **Ans**

Worked Example 6.2

Q   A four-pole armature has 600 lap-connected conductors, and is driven at 1500 rev/ min. Calculate the useful flux per pole required to generate an emf of 250 V.

A

$p = 2$; $z = 600$; $n = 1500/60$ rev/s; $a = 4$

$$E = \frac{2p\Phi zn}{a} \text{ volt}$$

therefore, $\Phi = \dfrac{Ea}{2pzn}$ weber $= \dfrac{250 \times 4 \times 60}{2 \times 2 \times 600 \times 1500}$

hence, $\Phi = 16.67$ mWb **Ans**

Worked Example 6.3

Q   A 4-pole d.c. generator is wound with 400 armature conductors and, under operating conditions, the emf generated in and current flowing through each conductor is 1.1 V and 40 A respectively. Calculate the total machine emf and generated power if the armature is (a) lap-wound, and (b) wave-wound.

A

$p = 2$; $z = 400$; $e = 1.1$ V; $i = 40$ A

(a) For a lap-wound machine, the number of parallel paths through the armature is equal to the number of poles, so $a = 4$. Thus, the total emf generated will be that generated by one quarter of the conductors in series.

$$E = \frac{ze}{a} \text{ volt} = \frac{400 \times 1.1}{4}$$
$$E = 110 \text{ V } \textbf{Ans}$$

Similarly, the total machine current will be the sum of the currents in the four parallel paths, so

$I = 4i = 4 \times 40$

$I = 160$ A **Ans**

$P = EI$ watt $= 110 \times 160$

$P = 17.6$ kW **Ans**

(b) For a wave-wound machine, the number of parallel paths, $a = 2$, regardless of the number of poles, so

$$E = \frac{400 \times 1.1}{2}$$
$$E = 220 \text{ V } \textbf{Ans}$$

and, $I = 2 \times 40$

$I = 80$ A **Ans**

$P = EI$ watt $= 220 \times 80$

$P = 17.6$ kW **Ans**

Worked Example 6.4

Q   **A 6-pole d.c. machine has a wave-wound armature of 500 conductors, and the useful flux/pole is 3 mWb. Determine the speed at which it must be driven in order to generate an emf of 240 V.**

A

$p = 3; a = 2; z = 500; \Phi = 3 \times 10^{-3}\,\text{Wb}; E = 240\text{V}$

$$E = \frac{2p\Phi zn}{a} \text{ volt}$$

$$\text{so, } n = \frac{Ea}{2p\Phi z} \text{ rev/second} = \frac{240 \times 2}{2 \times 3 \times 3 \times 10^{-3} \times 500}$$

$$= 53.33 \text{ rev/s}$$

$$= 53.33 \times 60 \text{ rev/min}$$

$$n = 3200 \text{ rev/min } \textbf{Ans}$$

6.7   Classification of Generators

D.C. generators are classified according to whether the field winding is electrically connected to the armature winding, and if it is, whether it is connected in parallel with or in series with the armature. The field current may also be referred to as the excitation current. If this current is supplied internally, by the armature, the machine is said to be self-excited. When the field current is supplied from an external d.c. source, the machine is said to be separately excited. The circuit symbol used for the field winding of a d.c. machine is simply the same as that used to represent any other form of winding. The armature is represented by a circle and two 'brushes'. The armature conductors, as such, are not shown.

6.8   Separately Excited Generator

The circuit diagram of a separately-excited generator is shown in Fig. 6.19. The rheostat, R_1, is included so that the field excitation current, I_f, can be varied. This diagram also shows the armature being driven at constant speed by some primemover. Since the armature of any generator must be driven, this drive is not normally shown. The load, R_L,

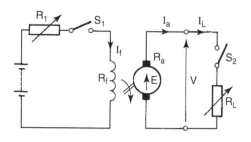

Fig. 6.19

being supplied by the generator may be connected or disconnected by switch S_2. The resistance of the armature circuit is represented by R_a.

Consider the generator being driven, with switches S_1 and S_2 both open. Let the armature terminals be monitored by a high impedance voltmeter (a DVM), such that negligible armature current flows. This meter will therefore indicate the generated emf. Despite the fact that there will be zero field current, a small emf would be measured. This emf is due to the small amount of residual magnetism retained in the poles. With switch S_1 now closed, the field current may be increased in discrete steps, and the corresponding values of generated emf noted. If the corresponding graph is plotted, it will be as shown in Fig. 6.20, and is known as the open-circuit characteristic of the machine.

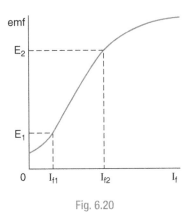

Fig. 6.20

It will be seen that the shape of this graph is similar to the magnetisation curve for a magnetic material. This is to be expected, since the emf will be directly proportional to the pole flux (refer back to the emf equation, and consider the terms p, z, a and n to be constants). The 'flattening' of the emf graph indicates the onset of saturation of the machine's magnetic circuit. When the machine is used in practice, the field current would normally be set to some value within the range indicated by I_{f1} and I_{f2} on the graph. This means that the facility exists to vary the emf between the limits E_1 and E_2 volts, simply by adjusting rheostat R_1.

Let the emf be set to some value E volt, within the range specified above. If the load is now varied, the corresponding values of terminal voltage, V and load current I_L may be measured. Note that with this machine the armature current is the same as the load current. The graph of V versus I_L is known as the output characteristic of the generator, and is shown in Fig. 6.21. The terminal p.d. of the machine will be less than the generated emf, by the amount of internal voltage drop due to R_a, such that:

$$V = E - I_a R_a \text{ volt} \tag{6.2}$$

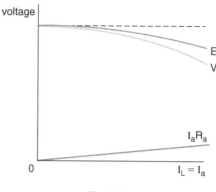

Fig. 6.21

Ideally, the graph of E versus I_L would be a horizontal line. However, the effects of armature reaction cause this graph to 'droop' at the higher values of current. The main advantage of this type of generator is that there is some scope for increasing the generated emf in order to offset the internal voltage drop, $I_a R_a$, as the load is increased. The big disadvantage is the necessity for a separate d.c. supply for the field excitation. The power flow diagram is shown in Fig. 6.22.

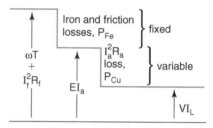

Fig. 6.22

6.9   Shunt Generator

This is a self-excited machine, where the field winding is connected in parallel (shunt) with the armature winding. The circuit diagram is shown in Fig. 6.23, and from this it may be seen that the armature has to supply current to both the load and the field, such that:

$$I_a = I_L + I_f \ \text{amp} \tag{6.3}$$

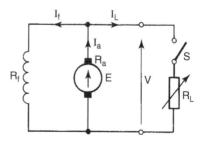

Fig. 6.23

Self-excitation process

This process can take place only if there is some residual flux in
the poles, and if the resistance of the field circuit is less than some
critical value. The open-circuit characteristic is illustrated in Fig.
6.24, where lines OA, OB and OC represent different values of shunt
field resistance. When the armature is rotated at constant speed, and
with switch S open, the initial emf generated will be E_1 volt, due to
the residual magnetism. This will cause a small current to circulate
through the field winding. This current will cause an increase in the
flux, and a consequent increase in emf. However, if the field resistance
(represented by OA) is too high, the generated emf will build up only
a small amount, to E_2 volt. Thus the machine has failed to self-excite.

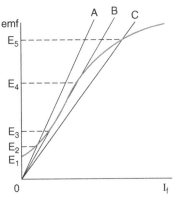

Fig. 6.24

Line OB represents the critical value of field resistance, referred to
above. With this value of resistance the machine emf would build up
to some value within the limits E_3 to E_4 volt. The machine has been
able to self-excite. However, the emf is in a highly unstable state,
since any slight disturbance, such as a fluctuation in speed, could
cause the emf to vary rapidly between the above limits. Thus the field
resistance needs to be less than this critical value as represented by OC.
Considering this line OC, it may be seen that at all points below E_5
volt, the generated emf is more than that required to merely maintain
I_f constant. Hence, more current flows, more flux produced, and more
emf is generated. Once the emf reaches E_5 volt, this is sufficient only
to maintain I_f constant, so the machine emf stabilises at this value.

The resistance of the field winding, R_f, is constant and of a relatively
high value compared with R_a. Typically, I_f will be in the order of 1A
to 10 A, and will remain reasonably constant. The shunt machine is
therefore considered to be a constant-flux machine. When switch S
is closed, the armature current will increase in order to supply the
demanded load current, I_L. Thus $I_a \propto I_L$, and as the load current is
increased, so the terminal voltage will fall, according to the equation

$V = E - I_a R_a$ volt. The output characteristic will therefore follow much the same shape as that for the separately excited generator. This condition applies until the machine is providing its rated full-load output. This means that (ignoring any losses) all of the mechanical input power, ωT, is being converted into electrical output power, VI_L watt. If the load should now demand even more current, this will be supplied by the machine, but only at the expense of field current. Thus, part of I_f is diverted to the load. This means that the flux will be reduced; E is reduced; V is reduced; I_L tries to increase to maintain the output power demand; I_f is further reduced; and so on. The result is that if the generator is overloaded, then it simply stops generating. This effect is shown by the dotted lines in Fig. 6.25, the output characteristics.

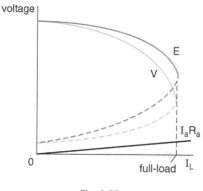

Fig. 6.25

The shunt generator is the most commonly used d.c. generator, since it provides a reasonably constant output voltage over its normal operating range. Its other obvious advantage is the fact that it is self-exciting, and therefore requires only some mechanical means of driving the armature. Figure 6.26 shows the power flow diagram.

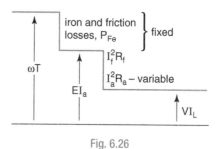

Fig. 6.26

6.10   Series Generator

In this machine the field winding is connected in series with the armature winding and the load, as shown in Fig. 6.27. In this case, $I_L = I_a = I_f$, so this is a variable-flux machine. Since the field winding

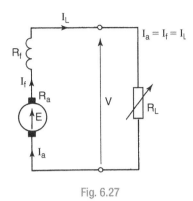

Fig. 6.27

must be capable of carrying the full-load current (which could be in hundreds of amps for a large machine), it is usually made from a few turns of heavy gauge wire or even copper strip. This also has the advantage of offering a very low resistance. This generator is a self-excited machine, provided that it is connected to a load when started. Note that a shunt generator will self-excite only when *disconnected* from its load.

When the load on a series generator is increased, the flux produced will increase, in almost direct proportion. The generated emf will therefore increase with the demanded load. The increase of flux, and hence voltage, will continue until the onset of magnetic saturation, as shown in the output characteristic of Fig. 6.28. The terminal voltage is related to the emf by the equation:

$$V = E - I_a(R_a + R_f) \text{ volt} \qquad (6.4)$$

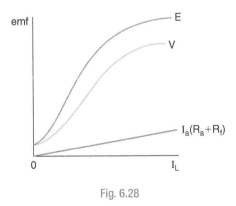

Fig. 6.28

The variation of terminal voltage with load is not normally a requirement for a generator, so this form of machine is seldom used. However, the rising voltage characteristic of a series-connected field winding is put to good use in the compound machine.

6.11   Compound Generator

This machine contains both shunt and series field windings, wound on to the same poles. The compound generator therefore combines the characteristics of both the shunt and series generators. The effect of the series field winding is designed to be relatively weak compared with that of the shunt field. In addition, the series field winding may be connected so as to either assist (strengthen) the shunt field, or to oppose (weaken) it. The former connection is referred to as *cumulative* compounding, and the latter as *differential* compounding. The circuit diagram for a compound generator is shown in Fig. 6.29. From this diagram it may be seen that the shunt field may be connected on either the armature side or load side of the series field winding. Since the shunt field is the predominant one, it is normally connected directly to the source from which it is supplied. Thus in the case of a generator it is connected directly to the armature, and this is called a *short shunt* connection. In the case of a motor, the source of supply is applied to the machine terminals, so the connection shown dotted in the circuit diagram is used. This is referred to as a *long shunt* connection.

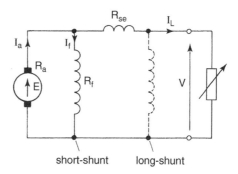

Fig. 6.29

Cumulative compounding

With this connection, the rising characteristic of the series field is used to compensate for the gradually falling characteristic of the shunt connection. Depending upon the relative strength of the series field, the machine may be either level compounded or slightly over compounded, as shown in Fig. 6.30. Level compounding is used for applications where excellent voltage regulation is required, i.e. terminal voltage on full-load is the same as that on no-load. If the electrical power output is to be transmitted some considerable distance, then the machine could be over compounded. The increase of terminal voltage with load would then compensate for the increased voltage drop along the transmission lines. If the degree of overcompounding is correctly chosen, then the voltage at the receiving end will remain constant, from no-load to full-load.

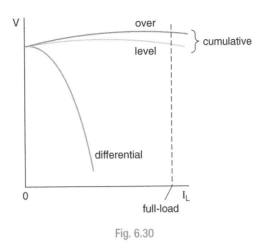

Fig. 6.30

Differential compounding

To achieve this, the connections to the series winding would be reversed, so that the current flowing through it will produce a flux in direct opposition to the shunt field flux. The total machine flux will therefore be the *difference* between the two fields. The consequence of this is that, as the load is increased, so the strength of the series flux is increased, but the overall machine flux is *decreased*. The result is the sharply falling voltage characteristic shown in Fig. 6.30. There are only limited applications for this type of characteristic, one of which is the supply to electric arc welding equipment. In this application, a relatively high voltage is required to strike the arc, but a comparatively lower voltage is sufficient to maintain the arc current.

The power flow diagram for a compound generator is shown in Fig. 6.31.

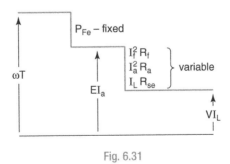

Fig. 6.31

Worked Example 6.5

Q   The resistance of the field winding of a shunt generator is 200 Ω. When the machine is delivering 80 kW the generated emf and terminal voltage are 475 V and 450 V respectively. Calculate (a) the armature resistance, and (b) the value of generated emf when the output is 50 kW, the terminal voltage then being 460 V.

A

$R_f = 200\ \Omega; P_o = 80 \times 10^3$ watt; $V = 450$ V; $E = 475$ V

The circuit diagram is shown in Fig. 6.32. It is always good practice to sketch the appropriate circuit diagram when solving machine problems.

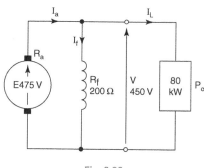

Fig. 6.32

(a) $P_o = VI_L$ watt; so $I_L = \dfrac{P_o}{V}$ amp

therefore $I_L = \dfrac{80 \times 10^3}{450} = 177.8$ A

$I_f = \dfrac{V}{R_f}$ amp $= \dfrac{450}{200} = 2.25$ A

$I_a = I_L + I_f$ amp $= 180.05$ A

$I_a R_a = E - V$ volt $= 475 - 450 = 25$ V

therefore $R_a = \dfrac{25}{180.05}$ ohm $= 0.139\ \Omega$ **Ans**

(b)   When $P_o = 50 \times 10^3$ W, $V = 460$ V

thus $I_L = \dfrac{50 \times 10^3}{460} = 108.7$ A

$I_f = \dfrac{V}{R_f} = \dfrac{460}{200} = 2.3$ A

hence, $I_a = 108.7 + 2.3 = 111$ A

$E = V + I_a R_a$ volt $= 460 + (111 \times 0.139)$

therefore $E = 475.4$ V **Ans**

Note: Although the load had changed by about 60%, the field current has changed by only about 2.2%. This justifies the statement that a shunt generator is considered to be a constant-flux machine.

Worked Example 6.6

Q   A short-shunt compound generator has armature, shunt-field and series-field resistances of 0.75 Ω, 100 Ω and 0.5 Ω respectively. When supplying a load of 5 kW at a terminal voltage of 250 V, calculate the generated emf.

A

$R_a = 0.75\ \Omega;\ R_f = 100\ \Omega;\ R_{Se} = 0.5\ \Omega;\ P_o = 5 \times 10^3\,W;\ V = 250\,V$

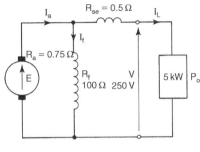

Fig. 6.33

$$I_L = \frac{P_o}{V}\ \text{amp} = \frac{5000}{250} = 20\ A$$

$$V_f = V + I_L R_{Se}\ \text{volt} = 250 + (20 \times 0.5)$$

so $V_f = 260\ V$

$$I_f = \frac{V_f}{R_f}\ \text{amp} = \frac{260}{100} = 2.6\ A$$

$$I_a = I_L + I_f\ \text{amp} = 22.6\ A$$

$$E = V_f + I_a R_a\ \text{volt} = 260 + (22.6 \times 0.75)$$

hence $E = 276.95\ V$ **Ans** (say 277 V)

Worked Example 6.7

Q   A d.c. shunt generator has armature and field resistances of 0.1 Ω and 180 Ω respectively. When the machine is supplying an output of 75 kW at a terminal voltage of 360 V and an efficiency of 86%, determine (a) the generated emf, (b) the variable loss, (c) the fixed losses, and (d) the iron, friction and windage loss.

A

$R_a = 0.1\ \Omega;\ R_f = 180\ \Omega;\ P_0 = 75 \times 10^3\ W;\ V = 360\,V;\ \eta = 86\% = 0.86$

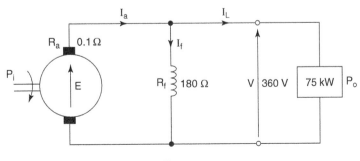

Fig. 6.34

(a) $P_o = V_L I_L$ watt

$I_L = \dfrac{P_a}{V_L}$ amp $= \dfrac{75 \times 10^3}{360}$

$I_L = 208.33$ A

$I_f = \dfrac{V}{R_f}$ amp $= \dfrac{360}{160} = 2$ A

$I_a = I_L + I_f$ amp $= 210.33$ A

$E = V + I_a R_a$ volt $= 360 + (210.33 \times 0.1)$

$E = 381$ V **Ans**

(b) From the power flow diagram of Fig. 6.26

Variable loss $= I_a^2 R_a$ watt $= 210.33^2 \times 0.1$

variable loss $= 4.424$ kW **Ans**

(c) $P_i = \dfrac{P_o}{\eta}$ watt $= \dfrac{75}{0.86}$ kW $= 87.21$ kW

total losses $= P_i - P_o$ watt $= 87.21 - 75$ kW

total losses $= 12.21$ kW

fixed losses $=$ total losses $-$ variable losses

$= 12.21 - 4.424$

fixed losses $= 7.785$ kW **Ans**

(d) $P_{Fe} =$ fixed losses $- I_f^2 R_f$ watt

$= 7785 - (2^2 \times 180)$

$P_{Fe} = 7.065$ kW **Ans**

6.12   D.C. Motors

All of the d.c. generators so far described could be operated as motors, provided that they were connected to an appropriate d.c. supply. Thus, the effects and problems associated with commutation and armature reaction apply equally to motors. Similarly the type of armature winding will have the same effect on the emf produced in the armature. When the machine is used as a motor, the armature generated emf is referred to as the back-emf, E_b.

Back-emf equation

As the armature of a motor rotates, an emf is induced in the armature winding, due to its motion through the magnetic field. The value of this back-emf will be exactly the same as would be generated if the machine was driven as a generator at the same speed. Thus, equation (6.1) could be expressed as follows:

$$E_b = \frac{2p\Phi zn}{a} \text{ volt} \tag{6.5}$$

The relationship between the back-emf and the supply voltage applied to the motor terminals would be a modified version of equation (6.2), such that:

$$E_b = V - I_a R_a \text{ volt} \qquad (6.6)$$

Note: For a generator, the emf must always be greater than the resulting terminal voltage. It must be borne in mind that for a motor to operate, current must be fed *into* it. Thus the supply voltage, *V*, applied to the machine terminals must be *greater than* the back-emf, E_b, in order to force current into the machine.

Speed equation

Considering equation (6.5), for any given machine, the terms $2p$, a and z are all constants. Thus we can say that:

$$E_b \propto \Phi n$$

$$\text{so, } n \propto \frac{E_b}{\Phi} \text{ or, } \omega \propto \frac{E_b}{\Phi} \qquad (6.7)$$

On examination, this relationship appears to be rather a strange result, since for a given value of back-emf, the *weaker* the flux, the *faster* the resulting speed of rotation. However, this fact is borne out in practice.

Torque equation

If some of the machine losses are ignored for a moment, then we can say that the mechanical power output is equal to the electrical power 'generated' in the armature, thus:

$$\omega T = E_b I_a$$
$$\text{so, } T = \frac{E_b I_a}{\omega}$$

and substituting for ω, from equation (6.7):

$$T \propto \frac{E_b I_a}{E_b / \Phi}$$

$$\text{therefore, } T \propto \Phi I_a \qquad (6.8)$$

The three equations (6.6), (6.7) and (6.8) may now be used to deduce the speed and torque characteristics for the various types of d.c. motor.

6.13   Shunt Motor

When the machine reaches its normal operating temperature, R_f will remain constant. Since the field winding is connected directly to a fixed supply voltage, V volt, then I_f will be fixed. Thus, the shunt motor (Fig. 6.35) is a constant-flux machine. Using the speed and torque equations we can say that:

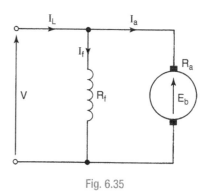

Fig. 6.35

Since $\omega \propto E_b / \Phi$, and Φ is constant; then

$$\omega \propto E_b \dots\dots\dots\dots[1]$$

Similarly, since $T \propto I_a\Phi$; then

$$T \propto I_a \dots\dots\dots\dots[2]$$

As the back-emf will have the same shape graph as that for the generator emf, and using [1] and [2] above the graphs of speed and torque versus current will be as in Fig. 6.36. Note that when the machine is used as a motor, the supply current is identified as I_L. In this case, the subscript 'L' represents the word 'line'. Thus I_L identifies the line current drawn from the supply, and I_a is directly proportional to I_L.

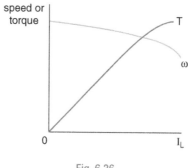

Fig. 6.36

Shunt motors are used for applications where a reasonably constant speed is required, between no-load and full-load conditions. The power flow diagram for any d.c. motor is, effectively, the power flow diagram of the corresponding generator, but turning this diagram 'back-to-front'. This is illustrated in Fig. 6.37, which is for a shunt motor. Compare this with Fig. 6.26.

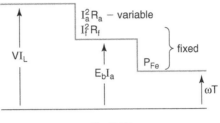

Fig. 6.37

6.14   Series Motor

Like the series generator, this machine is a variable-flux machine. Despite this, the back-emf of this motor remains almost constant, from light-load to full-load conditions. This fact is best illustrated by considering the circuit diagram (Fig. 6.38), the back-emf equation, and some typical values.

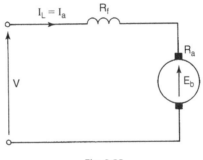

Fig. 6.38

$$E_b = V - I_a(R_a + R_f) \text{ volt} \qquad (6.9)$$

Let us assume the following: $V = 200$ V; $R_a = 0.15\,\Omega$; $R_f = 0.03\,\Omega$; $I_a = 5$ A on light-load; $I_a = 50$ A on full-load

Light-load: $E_b = 200 - 5(0.15 + 0.03) = 199.1$ V
Full-load: $E_b = 200 - 50(0.15 + 0.03) = 191$ V

From the above figures, it may be seen that although the armature current has increased tenfold, the back-emf has decreased by only 4%. Hence, E_b, remains sensibly constant.

Since $\omega \propto E_b/\Phi$, and E_b is constant, then:

$$\omega \propto \frac{1}{\Phi} \dots\dots\dots\dots[1]$$

Similarly, $T \propto \Phi I_a$, and since $\Phi \propto I_a$ until the onset of magnetic saturation,

then $T \propto I_a^2$.....................................[2]

and after saturation, $T \propto I_a$.................[3]

Using [1] to [3] above, the speed and torque characteristics shown in Fig. 6.39 may be deduced.

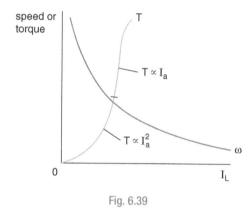

Fig. 6.39

Note: From the speed characteristic it is clear that, on very light loads, the motor speed would be excessive. *Theoretically,* the no-load speed would be infinite! For this reason a series motor must NEVER be started unless it is connected to a mechanical load sufficient to prevent a dangerously high speed. Similarly, a series motor must not be used to operate belt-driven machinery, lifting cranes etc., due to the possibility of the load being suddenly disconnected. If a series motor is allowed to run on a very light load, its speed builds up very quickly. The probable outcome of this is the distintegration of the machine, with the consequent dangers to personnel and plant.

The series motor has a high starting torque due to the 'square-law' response of the torque characteristic. For this reason, it tends to be used mainly for traction purposes. For example, an electric train engine requires a very large starting torque in order to overcome the massive inertia of a stationary train. The power flow diagram appears in Fig. 6.40.

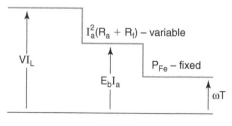

Fig. 6.40

6.15 Compound Motors

In order to make maximum use of the supply voltage, the long-shunt configuration is favoured. This is shown in Fig. 6.41. The series winding may, once more, be connected so that its flux either assists or opposes the flux from the shunt winding. The strength of the series field is kept weak compared with the shunt field.

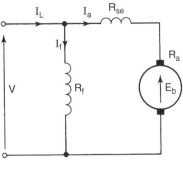

Fig. 6.41

Differential compounding

As the load on the machine is increased, the machine draws more current from the supply, and the armature current increases proportionately. Thus the strength of the series field is increased. However, the shunt field is constant, so the overall machine flux is reduced. The result is that the machine will tend to speed up. When the load on a machine is increased, its natural tendency is to slow down. Thus, if the machine is level compounded, this tendency for a drop in speed can be compensated, so that the full-load and no-load speed is the same. The effect of the shunt field is to limit the no-load speed to a safe value. However, the speed will tend to become too high if the machine is overloaded. The differentially compounded motor is used when excellent speed regulation is required. The speed characteristic is shown in Fig. 6.42.

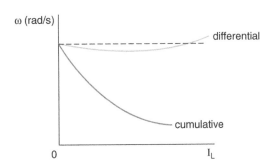

Fig. 6.42

Cumulative compounding

Again, the shunt winding is sufficient to limit the no-load speed to a safe value. As the load is increased, so too is the strength of the series field. Since the fluxes from the two fields are added, the overall machine flux also increases, with a consequent reduction in speed. The machine speed therefore drops considerably as the load is increased. A typical application requiring this type of characteristic would be an electrically driven press.

Figure 6.43 shows the power flow diagram for a compound motor.

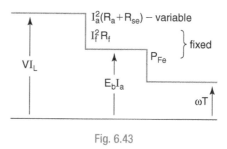

Fig. 6.43

6.16   Separately Excited Motor

This machine will have similar characteristics to those of the shunt motor. Its main advantage is the relative ease with which the speed may be varied, by controlling the field current. The direction of rotation can also be simply changed by reversing the field current. This type of motor is frequently used as a d.c. servo motor, particularly in its split-field form.

Worked Example 6.8

Q   A d.c. shunt motor develops a torque of 30 Nm at a speed of 850 rev/min when running from a 450 V supply and taking an armature current of 25 A. Calculate its speed and torque when running from a 220 V supply and taking an armature current of 12 A. The armature resistance is 0.8 Ω, and the flux at 220 V is 80% of that at 450 V.

A

$T_1 = 30$ Nm; $n_1 = 860$ rev/min; $V_1 = 450$ V; $I_{a1} = 25$ A;
$V_2 = 220$ V; $I_{a2} = 12$ A; $\Phi_2 = 0.8\ \Phi_1$ weber

In general, $n \propto \dfrac{E_b}{\Phi}$;   where $E_b = V - I_a R_a$ volt

so, $E_{b1} = 450 - (25 \times 0.8) = 430$ V

and, $E_{b2} = 220 - (12 \times 0.8) = 210.4$ V

$$n_1 \propto \frac{E_{b1}}{\Phi_1} \quad\dots\dots\dots\dots\dots [1]$$

$$\text{and } n_2 \propto \frac{E_{b2}}{\Phi_2} \quad\dots\dots\dots\dots [2]$$

and dividing [2] by [1]:

$$\frac{n_2}{n_1} = \frac{E_{b2}\Phi_1}{E_{b1}\Phi_2} = \frac{210.4 \times \Phi_1}{430 \times 0.8 \times \Phi_1} = 0.6116$$

$$\text{so, } n_2 = 0.6116 \times n_1 = 0.6116 \times 850$$

therefore $n_2 = 520$ rev/min **Ans**

Since $T \propto \Phi I_a$, then $T_1 \propto \Phi_1 I_{a1}$ and $T_2 \propto \Phi_2 I_{a2}$

$$\text{thus, } \frac{T_2}{T_1} = \frac{\Phi_2 I_{a2}}{\Phi_1 I_{a1}} = \frac{0.8 \times \Phi_1 \times 12}{\Phi_1 \times 25}$$

$$T_2 = 0.384 \times 30$$

$$\text{so } T_2 = 11.52 \text{ Nm } \textbf{Ans}$$

Worked Example 6.9

Q   A 4-pole d.c. motor draws an armature current of 85 A from a 400 V supply. The armature resistance is 0.2 Ω and is lap-wound with 800 conductors. If the useful flux/pole is 45 mWb, calculate (a) the speed of rotation, (b) the gross torque developed by the armature, and (c) the shaft output torque if the iron, friction and windage loss is 2.1 kW.

A

$p = 2; I_a = 85$ A; $V = 400$V; $R_a = 0.2\,\Omega; z = 800; a = 4; \Phi = 0.045$ Wb; $P_{Fe} = 2100$ W

(a) $E_b = V - I_a R_a$ volt $= 400 - (85 \times 0.2)$

$$E_b = 283 \text{ V}$$

$$E_b = \frac{2p\Phi zn}{a} \text{ volt}$$

$$\text{so, } n = \frac{E_b a}{2p\Phi z} \text{ rev/second} = \frac{383 \times 4}{4 \times 0.045 \times 800}$$

$$n = 10.64 \text{ rev/s}$$

$$n = 638 \text{ rev/min } \textbf{Ans}$$

(b) $\omega = 2\pi n$ radian/second $= 2\pi \times 10.64$

$$\omega = 66.85 \text{ rad/s}$$

gross 'generated' power $= \omega T_a = E_b I_a$ watt

$$T_a = \frac{E_b I_a}{\omega} \text{ newton metre} = \frac{383 \times 85}{66.85}$$

$$T_a = 487 \text{ Nm} \textbf{Ans}$$

(c) The difference between the shaft output power and the gross 'generated' power is the iron, friction and windage loss, P_{Fe}

$$\text{so, } P_o = E_b I_a - P_{Fe} \text{ watt} = (383 \times 85) - 2100$$

$$P_o = 30\,455 \text{ W}$$

and, $P_o = \omega T$ watt, where T is the shaft output torque

$$\text{so, } T = \frac{P_o}{\omega} \text{ newton metre} = \frac{30\,455}{66.85}$$

$$T = 455.6 \text{ Nm } \textbf{Ans}$$

Worked Example 6.10

Q   A 120 kW, 200 V shunt generator having a field resistance of 40 Ω was tested and yielded the following results:

(i) When running light as a motor, at full speed, the line current drawn from a 200 V supply was 30 A, and the field current was 5 A.

(ii) With the machine stationary, an applied voltage of 2 V across the armature caused a current of 15 A to flow.

Using these results, determine the machine's efficiency when operated as a generator on (a) full-load, and (b) half full-load. You may assume that the flux remains constant.

A

Full-load $P_o = 120 \times 10^3$ W; $V = 200$ V; $R_f = 40 \, \Omega$
as a motor on no-load: $I_L = 30$ A; $I_f = 5$ A
static armature d.c. test: $V_a = 2$ V; $I_a = 15$ A
from the motor test, fixed losses $= VI_L$ watt $= 200 \times 30$
fixed losses $= 6000$ W

from static test, $R_a = \dfrac{V_a}{I_a}$ ohm $= \dfrac{2}{15} = 0.133 \, \Omega$

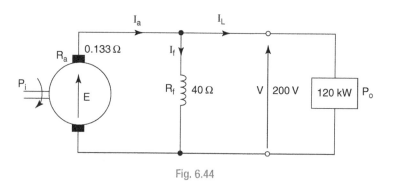

Fig. 6.44

(a) $I_L = \dfrac{P_o}{V}$ amp $= \dfrac{120 \times 10^3}{200}$

$I_L = 600$ A

$I_f = 5$ A

$I_a = I_L + I_f$ amp $= 605$ A

var. loss $= I_a^2 R_a$ watt $= 605^2 \times 0.133$
var. loss $= 48.681$ kW
total losses $= 48.681 + 6 \, \text{kW} = 54.681$ kW

$P_i = P_o + \text{losses} = 120 + 54.681$
$P_i = 174.681$ kW

$\eta = \dfrac{P_o}{P_i} \times 100\% = \dfrac{120}{174.68} \times 100\%$

$\eta = 68.7\%$ **Ans**

(b) On half full-load, $I_L = 300$ A, and $I_f = 5$ A

so $I_a = 305$ A

and $I_a^2 R_a = 305^2 \times 0.133 = 12.372$ kW

total loss $= 12.372 + 6 = 18.372$ kW

so $\eta = \dfrac{120}{138.372} \times 100\%$

$\eta = 86.7\%$ **Ans**

6.17   The Importance of Back-emf

We have seen that the resistance of the armature of a d.c. machine is very small, usually considerably less than one ohm. On the other hand, the supply voltage is often in hundreds of volts. If there was no back-emf to limit the armature current, the latter would be excessive. For example, if $V = 200$ V and $R_a = 0.5\,\Omega$, then in the absence of a back-emf, $I_a = 400$ A. This value of current would almost certainly cause severe damage to the armature winding. Now, whilst the armature is rotating, a back-emf will be generated, so the above problem will not exist. However, at the instant of connection to the supply the armature *will* be stationary, and an excessive current would be demanded from the supply. For this reason, some means of limiting the initial starting current must be employed. This effect is achieved by connecting extra resistance in series with the armature. As the machine speed builds up and the back-emf starts to increase, this extra resistance is gradually reduced to zero. The device used for this purpose is the faceplate starter.

6.18   D.C. Faceplate Starter

The general arrangement of the starter, connected between the supply and a shunt motor, is illustrated in Fig. 6.45. The arm is spring-loaded to the 'OFF' position. On the underside of the arm are two plates which make contact with a set of copper studs and a copper strip, when the arm is moved towards the 'ON' position. On the side of the arm is a soft-iron plate which makes contact with the poles of the no-volt relay (NVR) when the arm is in the 'ON' position. Connected between the studs are resistance elements. The supply to the arm contact plates is via the coil of the overload relay (OLR). The starting procedure is as follows:

The arm is moved so that the supply is connected to the first stud. Thus the armature is supplied via the whole of the starter resistance. The field winding is supplied directly via the copper strip and the coil of the NVR. The armature therefore starts to rotate, and its speed will build up to some steady value. As the arm is moved from stud to stud, the extra resistance in the armature circuit is reduced, and the speed increases. When the last stud is reached, the armature is connected

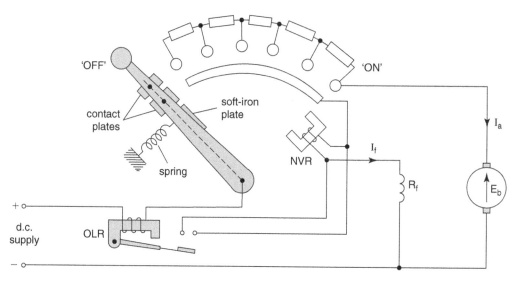

Fig. 6.45

directly to the supply, and the machine will be running at its normal no-load speed. With the arm in the 'ON' position, the flux produced by the coil of the NVR will be sufficient to hold the arm in this position, by attracting the soft-iron plate.

The NVR, the OLR and the spring on the arm provide protection for the motor, as follows. If the supply should fail whilst the machine is running normally, the coil of the NVR will de-energise, thus releasing the plate, and the spring returns the arm to the 'OFF' position. This procedure is necessary, otherwise, if the arm remained in the 'ON' position, when the supply is restored the stationary armature would be connected directly to the supply. The NVR therefore ensures that the correct starting procedure must be repeated after failure of the supply. The OLR has a hinged armature which carries an electrical contact. Whilst the line current from the supply is equal to or less than 1.5 times the normal full-load value, the flux produced by the OLR coil is insufficient to pull-in its armature. If the machine is seriously overloaded then the OLR armature will close, thus short-circuiting the NVR coil. The result is that the NVR de-energises and the spring returns the arm to the 'OFF' position.

6.19   Efficiency of D.C. Machines

As with any machine, the efficiency may be calculated from the ratio of the output power to the input power. In addition, the losses involved can be split between those that are unvarying (fixed losses) and those that will vary according to the loading placed on the machine. The condition for maximum efficiency is again achieved when the variable losses are equal to the fixed losses. The power flow diagram for a given type of machine identifies this division of losses. These diagrams can prove very useful when solving machine problems, and it is recommended that they be referred to when calculations are undertaken.

Provided that the armature and field circuit resistances are known, then the copper losses, P_{C_u}, can be easily calculated for given load conditions. The iron, and friction losses, P_{Fe}, are determined by conducting a simple test, as follows:

The machine is run light (on no-load) as a motor, being supplied with its normally rated terminal voltage. Under these conditions the armature current will be very small compared to its normal full-load value. For this reason the variable losses are negligible, and the input power to the machine may be taken as the machine's fixed losses. For the different types of machine, normally used as either motors or generators, the above simple test yields the following:

$$\text{Shunt machine:} \qquad VI_L = P_{Fe} + I_f^2 R_f \ \text{watt}$$
$$\text{Series machine:} \qquad VI_L = P_{Fe} \ \text{watt}$$
$$\text{Compound machine:} \quad VI_L = P_{Fe} + I_f^2 R_f \ \text{watt}$$

Worked Example 6.11

Q   A 240 V shunt motor, running on no-load and at normal speed, takes an armature current of 2.4 A from a 240 V supply. The resistance of the field circuit is 240 Ω, and that of the armature is 0.25 Ω. Calculate the motor output power and efficiency when drawing a line current of 35 A from the supply.

A

From the data given regarding the no-load conditions, the machine's fixed losses may be determined as follows. The circuit for these conditions is shown in Fig. 6.46.

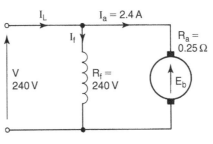

Fig. 6.46

$$\text{On no-load: } I_f = \frac{V}{R_f} \text{ amp} = \frac{240}{240} = 1 \text{ A}$$

$$I_L = I_a + I_f \text{ amp} = 2.4 + 1 = 3.4 \text{ A}$$

$$\text{fixed losses} = VI_L \text{ watt} = 240 \times 3.4$$

$$\text{so fixed losses} = 816 \text{ W}$$

The circuit conditions on load are shown in Fig. 6.47.

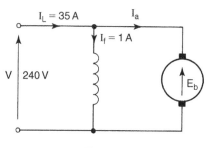

Fig. 6.47

$$\text{On load: } I_a = I_L - I_f \text{ amp} = 35 - 1 = 34 \text{ A}$$

$$P_{Cu} = I_a^2 R_a \text{ watt} = 34^2 \times 0.25 = 289 \text{ W}$$

$$\text{total losses} = 816 + 289 = 1105 \text{ W}$$

$$\text{input power, } P_i = VI_L \text{ watt} = 240 \times 35 = 8400 \text{ W}$$

$$\text{output, } P_o = P_i - \text{losses} = 8400 - 1105$$

$$\text{therefore, } P_o = 7.295 \text{ kW } \textbf{Ans}$$

$$\eta = \frac{P_o}{P_i} \times 100\% = \frac{7295}{8400} \times 100\%$$

$$\text{so } \eta = 86.85\% \textbf{ Ans}$$

Worked Example 6.12

Q   A shunt generator produces a full-load output of 10 kW at a terminal voltage of 250 V. The resistances of the armature and field circuits are 0.5 Ω and 125 Ω respectively. After running the machine as a motor on no-load, it was found that the iron and friction losses are 500 W. Determine (a) the power required at the driving shaft to provide the full-load output, (b) the input torque at full load if the driving speed is 900 rev/min, and (c) the full-load efficiency.

A

The circuit diagram is shown in Fig. 6.48.

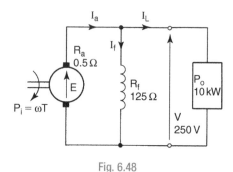

Fig. 6.48

(a) $I_L = \dfrac{P_o}{V}$ amp $= \dfrac{10^3}{250} = 40$ A

$I_f = \dfrac{V}{R_f}$ amp $= \dfrac{250}{125} = 2$ A

$I_a = I_L + I_f$ amp $= 42$ A*

$E = V + I_a R_a$ volt $= 250 + (42 \times 0.5) = 271$ V

generated power,

$EI_a = 271 \times 42 = 11\,382$ W

$P_i = EI_a + P_{Fe}$ watt $= 11\,382 + 500$

so $P_i = 11.882$ kW **Ans**

(b) $\omega = 2\pi n$ rad/s; where $n = 900/60$ rev/s

$= 2 \times \pi \times \dfrac{900}{60} = 94.25$ rad/s

$\omega T = P_i = 11\,882$ W

so $T = \dfrac{11\,882}{94.25} = 126.1$ Nm **Ans**

(c) $\eta = \dfrac{P_o}{P_i} \times 100\% = \dfrac{10\,000}{11\,882} \times 100$

hence $\eta = 84.16\%$ **Ans**

*alternatively, $P_{Cu} = I_a^2 R_a + I_f^2 R_f$ watt

$= (42^2 \times 0.5) + (2^2 \times 125)$

so $P_{Cu} = 1382$ W

total losses $= P_{Cu} + P_{Fe}$ watt $= 1382 + 500 = 1882$ W

$P_i = P_o +$ losses $= 11.882$ kW **Ans**

6.20   Simple Methods of Speed Control

The speed of a d.c. motor may be controlled by varying either the armature current or the field current. With self-excited machines, the only way these two currents can be varied is to reduce them from their full-load values. This is normally achieved by some form of variable resistance, connected in the relevant circuit.

Field regulator

This method is suitable for use only with shunt and compound machines. The regulator consists of a variable resistor connected in series with the shunt field winding. When this resistance is increased, the current decreases, thus weakening the flux. This results in an increase of speed. Using this method it is possible to increase the speed to three or four times that at full excitation.

Controller

This is a variable resistor connected in series with the armature, similar to starter resistance. As this resistance is increased, so the back-emf is reduced, resulting in a decrease of speed. Speeds from zero to full speed are possible, but this method is very wasteful of power, and hence severely reduces the efficiency of the motor.

Separate excitation

As stated in Section 6.16, the speed and direction of rotation of a separately excited motor is relatively easy to control, simply by varying the field current.

The speed of a series (or universal) motor used in small domestic appliances, such as a food mixer, is often achieved by switching different field coils into and out of circuit. This results in a predetermined speed for each field coil.

6.21   Stepper Motors

Stepper or stepping motors cannot readily be classified as either d.c. or a.c. machines. Logically they should be classified as digital machines, since they are controlled by electronic circuitry that in turn controls the sequence of pulses fed to the motor coils. The other main feature of this type of motor is that the rotation occurs in discrete angular steps. The step angle may be either large (say 30°) or small, depending upon the application. A step angle of 1.8° is quite common, which allows 200 steps per revolution. The stepping speed is controlled by the frequency or pulse rate of the control signal, and up to about 800 steps/second can be achieved. Since the angular position of the machine's rotor can be controlled with great precision, stepper motors are used in applications such as robotics, numerically controlled machines, head positioners for floppy and hard disc drives, X-Y plotters, video cassette recorder heads etc.

A simple form of stepper motor is illustrated in Fig. 6.49. The rotor is made from a 'soft' magnetic material, and carries no electrical

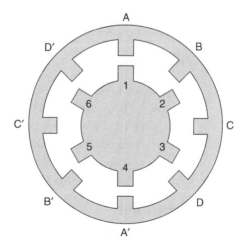

Fig. 6.49

windings. The number of rotor teeth will be different to the number of stator poles. In the case shown, the stator poles are displaced by 45°, whereas the rotor teeth are displaced by 60°. As will be shown, this will result in a step angle of 15°. Windings are mounted on the stator poles in such a way that they can produce magnetic flux in the axes AA′, BB′, CC′, etc. when they are energised from the control circuit.

Consider that stator coils AA′ have been energised, and the rotor has aligned with the resulting flux, as shown in the diagram. If AA′ is de-energised as BB′ is energised, the flux axis rotates through 45° clockwise. Rotor teeth 2 and 5 will now align with this flux, thus turning the rotor through a 15° step anticlockwise. If BB′ is then de-energised as CC′ is energised, rotor teeth 3 and 6 will align with CC′; and so on. The direction of rotation is simply reversed by reversing the sequence in which the stator coils are energised.

Summary of Equations

Generators:

EMF equation: $E = \dfrac{2p\Phi zn}{a}$ volt

for a lap winding, $a = 2p$; and for a wave winding, $a = 2$

Shunt generator: $I_a = I_L + I_f$ amp
$$V = E - I_a R_a \text{ volt}$$

Series generator: $I_a = I_L = I_f$ amp
$$V = E - I_a(R_a + R_f) \text{ volt}$$

Compound generator: $I_a = I_L + I_f$ volt
$$V = E - (I_a R_a + I_L R_{Se}) \text{ volt — for a short shunt}$$

Motors:

Back-emf equation: $E_b = \dfrac{2p\Phi zn}{a}$ volt

Shunt motor: $E_b = V - I_a R_a$ volt

Series motor: $E_b = V - I_a(R_a + R_f)$ volt

Compound motor: $E_b = V - I_a(R_a + R_{Se})$ volt — for a long shunt

Speed equation: $n \propto \dfrac{Eb}{\Phi}$ rev/second; or $\omega \propto \dfrac{Eb}{\Phi}$ rad/second

Torque equation: $T \propto \Phi I_a$ newton metre

Efficiency: $\eta = \dfrac{\text{output}}{\text{output} + \text{losses}} \times 100\%$

η_{max} occurs when variable losses = fixed losses

Assignment Questions

1   A four-pole wave-wound machine has 400 armature conductors and is driven at 900 rev/min. If the generated emf is 200 V, determine the useful flux per pole.

2   A six-pole machine has 420 armature conductors. If the flux per pole is 35 mWb, determine the speed of rotation required in order to generate an emf of 250 V when the armature is (a) lap-wound and (b) wave-wound.

3   A shunt generator supplies a current of 85 A at a terminal p.d. of 380 V. Calculate the generated emf if the armature and field resistances are 0.4 Ω and 95 Ω respectively.

4   A generator produces an armature current of 50 A when generating an emf of 400 V. If the terminal p.d. is 390 V, calculate (a) the value of the armature resistance, and (b) the power loss in the armature circuit.

5   A d.c. shunt generator supplies a 50 kW load at a terminal voltage of 250 V. The armature and field circuit resistances are 0.15 Ω and 50 Ω respectively. Calculate (a) the generated emf, and (b) the efficiency if the mechanical and iron losses total 2.45 kW.

6   A short-shunt compound generator has armature, shunt-field and series-field resistances of 0.8 Ω, 50 Ω and 0.6 Ω respectively. The shunt-field copper loss is 312.5 W; the series-field copper loss is 240 W, and the terminal voltage is 250 V. Determine (a) the power output, and (b) the generated emf.

7   A shunt-wound machine has armature and field circuit resistances of 0.05 Ω and 105 Ω respectively. When the machine is driven as a generator at 650 rev/min it produces an output of 50 kW at a terminal voltage of 420 V. Determine the speed at which it would run as a motor when taking an input of 50 kW from a 420 V d.c. supply.

8   A series motor runs at 600 rev/min when connected to a 240 V d.c. supply, and drawing 30 A. Calculate its speed when drawing a current of 42 A from a 300 V supply. The combined armature and field resistance is 0.48 Ω. You may assume that the flux remains directly proportional to the field current.

9   A four-pole motor has its armature lap-wound with 1000 conductors, and it runs at 1000 rev/min when taking an armature current of 40 A from a 250 V supply. Given that the armature resistance is 0.18 Ω, calculate (a) the useful flux per pole, and (b) the gross torque developed by the armature.

10   A shunt generator has a full-load output of 10 kW at a terminal voltage of 240 V. The armature and field resistances are 0.5 Ω and 150 Ω respectively. The mechanical and iron losses total 500 W. Determine (a) the input driving power required at full-load, (b) the full-load efficiency, and (c) the armature current that results in maximum efficiency.

11   A d.c. motor was tested by connecting its shaft to the brake pulley shown in Fig. 6.50. The suspended mass was 25 kg and the reading on the spring balance was 60 N when the motor was running at 950 rev/min. Given that the diameter of the brake pulley was 350 mm, and the motor input was 20 A at a terminal voltage of 200 V, calculate (a) the motor output, and (b) its efficiency at this load.

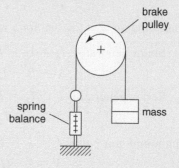

Fig. 6.50

12   A d.c. shunt generator has a full-load output of 180 kW at 250 V. The machine was tested as detailed below, yielding the results shown.

(i) When running light as a motor at full speed, it drew a line current of 35 A from a 250 V supply, and the field current was 10 A.

(ii) With the machine stationary, an emf of 3 V across the armature produced a current of 200 A.

Using these results, determine the full-load efficiency.

13   The current drawn by a 450 V shunt motor, when running light, is 8 A. The armature and field resistances are 0.15 Ω and 250 Ω respectively. Calculate (a) the input power and the efficiency when a current of 120 A is drawn from this 450 V supply, and (b) the armature current at which maximum efficiency occurs.

Suggested Practical Assignments

Assignment 1

To obtain the open-circuit and output characteristics for a separately excited d.c. generator.

Apparatus:

1 × separately excited generator (fhp machine)
1 × d.c. supply (e.g. 110 V)
2 × single-pole switch
2 × rheostats
2 × ammeter
1 × voltmeter
1 × constant-speed drive motor

Method:

1 Connect the circuit shown in Fig. 6.51, with both switches left open and both rheostats set to maximum resistance.

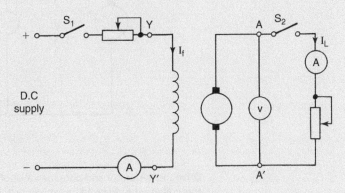

Fig. 6.51

2 Operate the drive motor, measure and tabulate any generated emf.
3 Close switch S_1, measure and tabulate the values of field current and generated emf.
4 Increase the field current, in discrete steps, and tabulate these current settings together with the corresponding emf generated. Continue this procedure until the generator is producing its normally rated emf.
5 Now close S_2, measure and tabulate the corresponding values of load current and terminal voltage.
6 Increase the load current, in discrete steps, tabulating the current and terminal voltage at each step. Continue this procedure up to the rated full-load output for the machine.
7 From your tabulated data, plot a graph of generated emf versus field current, and a graph of terminal voltage versus load current.
8 Complete an assignment report, including comments regarding the shapes of the two characteristics compared with those predicted by theory.

Assignment 2

To obtain the output characteristics for a shunt-wound d.c. generator.

Apparatus:

1 × shunt generator
1 × drive motor
1 × rheostat
1 × ammeter
1 × voltmeter
1 × single-pole switch

Method:

1   Connect the circuit as in Fig. 6.52, leaving the switch in the open position, and the rheostat set to its maximum resistance setting.

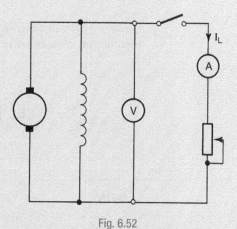

Fig. 6.52

2   Drive the generator and note the value of generated voltage. Note that this voltage should build up rapidly to its normally rated output value. If self-excitation does not occur, switch off the drive motor, reverse the connections to the generator shunt field, and restart the motor. If the generator still fails to self-excite, consult your lecturer.

3   Close the switch, note and tabulate the corresponding values of load current and terminal voltage.

4   Increase the load current, in steps, noting the values of load current and terminal voltage at each step. Continue this procedure up to the full-load output of the machine.

5   If possible, increase the load current beyond the full-load value, and observe the effects on both this current and the terminal voltage.

6   Prepare an assignment report. This should include a plotted graph of the generator output characteristic, and comments regarding its shape etc.

Assignment 3

To obtain the output characteristics for a compound generator, using both cumulative and differential compounding.

Apparatus:

$1 \times$ compound generator
Remainder of apparatus as for Assignment 2.

Method:

As for Assignment 2, except that the procedures will be carried out for both cumulative and differential compounding. In order to change from one form of compounding to the other, simply reverse the connections to the series field winding. The circuit is shown in Fig. 6.53.

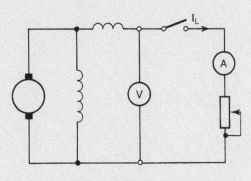

Fig. 6.53

Chapter 7

Measuring Instruments and Measurements

Learning Outcomes

This chapter is intended to help you understand the principal features and limitations of some commonly used measuring instruments. On completion you should be able to:

1 Appreciate the sources of error that can occur when taking measurements.
2 Predict the loading effects and frequency response characteristics of instruments.
3 Predict the effect of distorted waveforms on detectors used in electronic instruments.
4 Calculate the power consumed by instruments connected into circuits.
5 Appreciate the basic principles of a.c. bridge measurements.
6 Select and correctly use appropriate instruments for a variety of measurements.

7.1   Introduction

The operating principles of moving coil and rectifier moving coil multimeters; Wheatstone bridge; slidewire potentiometer; and CRO are covered in *Fundamental Electrical and Electronic Principles.* In addition, you will, by now, have gained some practical experience of simple voltage, current and resistance measurements, using the above instruments. As part of this experience you may well have come to the important realisation that *no* measurement should ever be considered as being absolutely correct. Any measurement that is made will involve a number of possible sources of error.

Calibration errors

The calibration of an instrument is the process whereby its indicated values are compared against a known standard. Since no instrument can be made perfect in all respects, there will be discrepancies between the indicated values and the 'true' values. These discrepancies are known as the calibration errors, and may be shown in the form of a graph, as in Fig. 7.1. There will be some allowable limits of error specified for

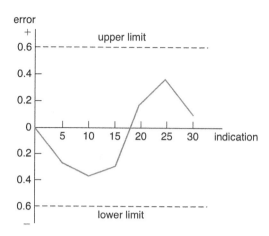

Fig. 7.1

the instrument. The range of these limits determines the 'grade' which the instrument is allocated; the lower the limits of error the 'higher' the grade. Figure 7.1 could represent the calibration graph for a 0 to 30 V voltmeter, with a maximum allowable error of ± 0.6 V. The accuracy of an instrument is always quoted in terms of the maximum allowable error, usually as a percentage of the full-scale deflection (fsd). Thus for the above example, the accuracy of the instrument would be quoted as being ±2% of fsd. It is worth noting that if the instrument was in error by±0.6 V at an indicated value of, say, 10 V, then the actual percentage error at this indication would be:

$$\pm\frac{0.6}{10}\times 100\% = \pm 6\%$$

This is because the allowable error of 0.6 V is a much more significant proportion of 10 V than it is of 30 V. For this reason, when using pointer-on-scale instruments, it is advisable to use an instrument (or the appropriate scale on the instrument) so that the indicated value is as near to fsd as is possible. When this is done, the calibration error is minimised. The effect of this form of error is illustrated in Example 7.1.

Worked Example 7.1

Q   The value of a 50 Ω resistor is to be verified by measuring the current flowing through it and the corresponding p.d. across it. Details of the two meters, and the readings recorded are shown below. Determine the range of values between which the actual resistor value may be said to lie.

Voltmeter: accuracy ± 2% of fsd; range 0−100 V; indicated reading 75 V

Ammeter: accuracy ± 2% of fsd; range 0−3 A; indicated reading 1.5 A.

A

Possible voltmeter error = ± 0.02 × 100 = ± 2V

therefore possible p.d. = 75 ± 2=73V to 77V

possible ammeter error $= \pm 0.02 \times 3 = \pm 0.06$ A

therefore possible current $= 1.5 \pm 0.06 = 1.44$ A to 1.56 A

Since $R = V/I$ ohm, then the maximum error of measurement will occur when one meter error is at its upper limit, and that for the other meter is at its lower limit.

i.e. when $V = 77$ V and $I = 1.44$ A

or when $V = 73$ V and $I = 1.56$ A

hence, $R = \dfrac{77}{1.44} = 53.472$ Ω;

or $R = \dfrac{73}{1.56} = 46.795$ Ω

Therefore, the resistor value may be said to have a value in the range 46.795 Ω and 53.472 Ω **Ans**

Systematic errors

This form of error is either avoidable, or may be taken into account when a measurement is made. Failure to ensure that an instrument is correctly 'zeroed' before connecting it to a circuit is a common but easily avoidable systematic error. The loading effect of an instrument, and the points in the circuit to which it is connected may also be taken into account in order to minimise this form of error. This form of error is illustrated in Example 7.2.

Worked Example 7.2

Q   A 5 kΩ resistor is connected across a 10 V supply. The p.d. across the resistor and the current flowing through it are measured as shown in Fig. 7.2. The voltmeter has a total resistance of 20 kΩ, and that of the ammeter is 5 Ω. Assuming negligible calibration errors, let us determine the systematic errors in the two meter readings, attributable to their manner of connection.

A

With neither meter connected, then the 'true' values for p.d. and current will be 10 V and 2 mA respectively.

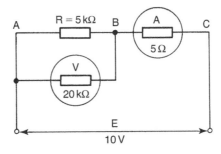

Fig. 7.2

$$R_{AB} = \frac{5 \times 20}{2 + 20} \text{ k}\Omega = 4 \text{ k}\Omega$$

$$R_{AC} = 4.005 \text{ k}\Omega; \text{ and}$$

$$V = \frac{R_{AB}}{R_{AC}} \times E = \frac{4}{4.005} \times 10 = 9.888 \text{ V}$$

This would be the p.d. across the voltmeter, and assuming that it could be accurately read down to three decimal places, then the error would be:

$$\frac{9.888 - 10}{10} \times 100\% = -1.12\%$$

This degree of error would be acceptable. Indeed, the voltmeter reading would most likely be interpreted as 10 V.

$$\text{The circuit current, } I = \frac{E}{R_{AC}} \text{ amp}$$

$$= \frac{10}{4005} = 2.497 \text{ mA}$$

This is the value of current measured by the ammeter, so the error in this reading would be:

$$\frac{2.497 - 2}{2} \times 100\% = +25\%$$

This degree of error is clearly unacceptable. However, if the voltmeter was connected between A and C, as in Fig. 7.3, then:

$$\text{Voltmeter reading} = 10 \text{ V, i.e. zero systematic error}$$

$$\text{Current through ammeter} = \frac{E}{R_{ABC}} \text{ amp} = \frac{10}{5005} = 1.998 \text{ mA}$$

$$\text{so ammeter error} = \frac{1.998 - 2}{2} \times 100\% = -0.1\%$$

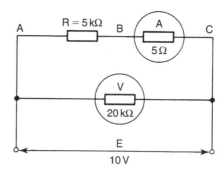

Fig. 7.3

This degree of error is most acceptable, so this alternative connection of the voltmeter is obviously better at reducing the total systematic error. It is left to the reader to verify that if the voltmeter resistance was (say)

$10\,\text{M}\Omega$, then the systematic error introduced by the voltmeter connection of Fig. 7.2 would be negligible. This clearly illustrates one advantage of a digital voltmeter compared with a moving coil instrument. That is, a perfect voltmeter would have an infinite resistance (draw zero current). On the other hand, a perfect ammeter would have zero resistance, and hence cause no additional potential drop in the circuit.

Observational and random errors

All other forms of measurement error may be included under this heading. Observational (human) error can occur in a number of ways. When using a pointer-on-scale instrument, the pointer often comes to rest between two marked graduations on the scale. The reading recorded is therefore open to interpretation by the individual observer, and different observers will tend to give slightly different interpretations. Occasionally there may be very large observational errors. This is fairly common when a student first uses a multimeter such as an AVO, and inadvertently reads the indication from the wrong scale, or misinterprets the scale setting applied by the rotary range switch. This problem with the range selection can also occur with digital instruments. With this form of instrument there can be no 'argument' as to the digits displayed, but the displayed value is normally subject to a ± 1 digit random error. Temperature and other environmental factors may also be the cause of random errors in measurements.

7.2   Accuracy and Sensitivity

These two measurement terms are often confused with each other, despite the fact that they describe different features of a measuring instrument or system.

Accuracy

This has already been described as the error referred to the fsd. Another way to define accuracy would be to say that it represents the closeness to the true value that is obtainable.

Sensitivity

The sensitivity, discrimination or resolution of an instrument describes the smallest change of the measured quantity (measurand) that can be discerned on the display. Thus sensitivity may be defined as:

$$\frac{\text{change of indication (output)}}{\text{change of measured quantity (input)}}$$

Comparing accuracy and sensitivity, it can be said that a sensitive instrument is not necessarily accurate; but an accurate instrument needs to be sensitive.

7.3   Total Measurement Error

The in-depth study of the theory of errors is beyond the scope of the course of study being undertaken here. However, a simplified coverage of the most common techniques will be outlined, in order to give an appreciation of the result of total error.

Addition and subtraction of measurements

Consider two meter readings which are to be added or subtracted to obtain a final result. Let D_1 and D_2 be the two displayed readings, and e_1 and e_2 the respective maximum errors. The sum of the two readings, D_s, and the total error is:

$$D_s = D_1 + D_2 \pm (e_1 + e_2)$$

and the difference,

$$D_d = D_1 - D_2 \pm (e_1 + e_2)$$

Hence the total error in the sum or difference of two or more readings is equal to the sum of the individual errors. Note that e_1 and e_2 are the absolute errors in the readings, and not the quoted accuracies of the instruments. For example, suppose that two p.d.s are measured in a series d.c. circuit, and the sum of the two represents the applied voltage. The voltmeter used has an accuracy of $\pm 2\%$ and an fsd of 100 V. If the two p.d.s indicated are $V_1 = 60\,$V and $V_2 = 20\,$V, the applied voltage and associated total error is obtained thus:

Since meter has an accuracy of $\pm 2\%$ of fsd, then absolute possible error is $\pm 2\,$V. The possible values for V_1 and V_2 are,

$$V_1 = 60 \pm 2 \text{ V and } V_2 = 20 \pm 2 \text{ V}$$
$$\text{so, } V = V_1 + V_2 = (60 + 20) \pm (2 + 2) \text{ V}$$
$$= 80 \pm 4 \text{ V}$$

Thus the total possible error is $\pm 4\,$V or $\pm 5\%$.

Multiplication and division of measurements

Consider two measurements and their associated errors as follows:

$$D_1 \pm e_1 = D_1\left(1 \pm \frac{e_1}{D_1}\right) = D_1(1 \pm \delta_1)$$

and similarly, $D_2(1 \pm \delta_2)$

where $\delta_1 = e_1/D_1$; $\delta_2 = e_2/D_2$ are the *fractional* errors. The product of the two readings yields the result,

$$D_p = D_1 D_2(1 \pm \delta_1)(1 \pm \delta_2)$$
$$= D_1 D_2[1 \pm (\delta_1 + \delta_2) + \delta_1 \delta_2]$$

but since δ is generally a very small quantity, then the sum $(\delta_1 + \delta_2)$ is normally very much greater than the product $\delta_1 \delta_2$.

$$\text{Therefore, } D_p = D_1 D_2[1 \pm (\delta_1 + \delta_2)]$$

So, for a product of two measurements, the total fractional error is the sum of the individual fractional errors. Similarly, it can be shown that when dividing one reading by another, the quotient,

$$D_q = \frac{D_1}{D_2}[1 \pm (\delta_1 + \delta_2)]$$

Power or root of a measurement

Since a root is simply a fractional power, then the same technique applies to both. For example, the square-root of a quantity expressed as a power is,

$$\sqrt{y} = y^{1/2}$$

It is found that the total fractional error in raising to a power n equals n times the original fractional error, such that:

reading raised to the power $n = D^n(1 \pm n\delta)$

7.4 Wattmeter Corrections

When a wattmeter is used to measure the power in a circuit, systematic errors are introduced. The amount of error thus introduced depends upon the way in which the wattmeter is connected into the circuit. Consider Fig. 7.4, which shows a wattmeter connected between a supply and its load. The wattmeter reading depends on the p.d. across its voltage coil, and the current flowing through its current coil. However, the current through the current coil is the load current, I_L, *plus* the small current I_p drawn by the voltage or pressure coil. The wattmeter reading is therefore:

$$P = V_L(I_L + I_p) \text{ watt}$$
$$= V_L I_L + V_L I_p$$
$$= \text{load power} + V_L I_p \text{ watt.}$$

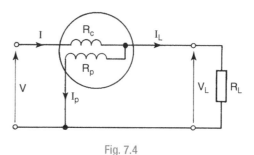

Fig. 7.4

The wattmeter reading is therefore too high, by the amount of power dissipated in the pressure coil. This coil of the wattmeter will have a relatively high resistance, so I_p will be correspondingly small. Thus, if $I_L \geqslant 10 \times I_p$, then the error introduced is negligible. If, on the other hand, the load current is of a comparable value to I_p, then the error can be significant. In this situation the wattmeter reading should be corrected by subtracting the pressure coil power from the meter reading. This is possible provided that the resistance of the pressure coil is known, such that:

$$P_p = \frac{V_L^2}{R_p} \text{ watt, or } P_p = I_p^2 R_p \text{ watt}$$

The wattmeter may be connected so that the pressure coil is connected to the supply side of the circuit as in Fig. 7.5. In this situation the current through the current coil will be the true load current. However, the p.d. across the pressure coil is the load p.d. plus the potential drop due to the resistance of the current coil. The total wattmeter reading is therefore:

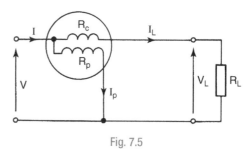

Fig. 7.5

$$P = (V_L + IR_c)I_L \text{ watt}$$
$$= V_L I_L + I_L^2 R_c$$
$$= \text{load power} + I_L^2 R_c$$

This means that the meter reading will again be high, and may be corrected provided that the current coil resistance is known. This form of connection will minimise the error when the load current (and hence the current coil p.d.) is relatively small.

7.5 Electronic Voltmeter

It has been shown that the main source of error for analogue voltmeters, such as the AVOmeter, occurs when measuring the p.d. across a high value resistor. This error is due to the loading effect of the meter's input resistance. Additionally, when used on the a.c. ranges the upper frequency limit will be in the order of 20 kHz. An electronic voltmeter reduces these sources of error by employing a transistor amplifier circuit. This amplifier is normally based on a field effect transistor (FET), which can have an input resistance in the order of hundreds of megohms. This amplifier may also be connected as a DC (directly coupled) amplifier, which enables it to have a bandwidth extending from 0 Hz up to many megahertz. Since this amplifier can have considerable voltage gain, there is a limitation on the size of input voltage that may be applied to it. For this reason a switchable attenuator network is interposed between the meter terminals and the amplifier input terminals. One disadvantage of this attenuator network is to reduce the overall input resistance of the instrument to the region of 10–20 MΩ. However, this is still far greater than that of an AVO, so the loading effect of the electronic voltmeter is minimal. Another advantage of employing an amplifier is that an electronic voltmeter may be capable of measuring voltages as low as microvolts. The amplified voltage is normally displayed on a conventional moving coil movement. For this reason, when the instrument is used for a.c. measurements, a rectifier is switched into the input circuit. A much simplified diagram of such an instrument is shown in Figs. 7.6(a) and (b).

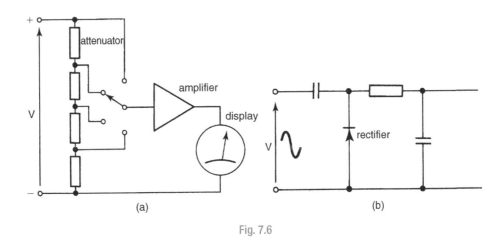

Fig. 7.6

7.6 Digital Voltmeter (DVM)

As with the electronic voltmeter, a DVM has a high input resistance because it employs a transistor amplifier and input attenuator. The amplifier is usually in the form of an operational amplifier connected as an integrator. This, coupled with an oscillator, a counter and associated logic control circuitry forms an analogue-to-digital converter (ADC) circuit. Due to the presence of frequency dependent components such as capacitors, the bandwidth of such an arrangement is much less than that achievable from an electronic voltmeter. Indeed, some of the commonly used (and 'cheaper') DVMs may have a bandwidth of only 2–4 kHz. It is therefore most important to consult the instrument data sheet to check on this aspect. Since an AVO has a better frequency response than the above figure, then this instrument could prove to be more reliable for a.c. measurements at frequencies greater than 2 kHz. For power frequency and d.c. measurements any DVM will normally be preferable to an analogue instrument.

7.7   Complex Waveforms

The simplest of waveforms is the sinewave. The most complex is a squarewave. If any complex waveform is analysed, it will be found that it is formed by the addition of a number of sinewaves, of different frequencies, amplitudes and phase relationships. The frequencies of these sinewaves are always whole integer multiples of the complex waveform frequency, i.e. 1, 2, 3, 4, 5, etc., times the frequency of the complex wave. The first of these has the same frequency as the complex wave, and is known as the *fundamental*. The fundamental normally has the largest amplitude of all of the sinewaves.

The remaining sinewaves are known as the *harmonics*. The sinewave of twice the fundamental frequency is called the second harmonic; the one with three times the frequency is the third harmonic, and so on. It is also usual to find that the higher the order of the harmonic, the smaller its amplitude. Thus, the fundamental has a smaller amplitude than the complex wave; the amplitude of the second harmonic is less than the fundamental, and so on. Some examples of the complex waveforms produced by the addition of the fundamental and various harmonics are illustrated in Figs. 7.7 (a) and (b).

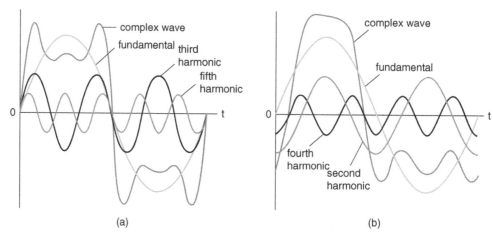

Fig. 7.7   (a) Odd harmonics 'in-phase', (b) even harmonics 'out-of-phase'.

There are several points worth noting about these two diagrams. Firstly, the references to 'in-phase' and 'out-of-phase'. Strictly speaking, we can only compare the phase relationship of waveforms of the *same frequency*. However, the harmonics are of different frequencies, so their phase relationships to each other are continuously changing. Hence, we can only compare their phase relationships at a given instant of time. In the diagrams shown, this instant has been chosen as time $t = 0$.

Considering Fig. 7.7(a), at this instant the fundamental and the other two harmonics are all passing through zero. Hence the reference to the harmonics being 'in-phase'. In Fig. 7.7(b), at $t = 0$, both of the even harmonics are effectively lagging the fundamental by 90°. Hence the reference to them being 'out-of-phase'.

Secondly, let us compare the shape of the two complex waves. Fig. 7.7(a) is a symmetrical waveform, whereas that shown in Fig. 7.7(b) is definitely non-symmetrical. In general it is found that odd harmonics tend to result in symmetrical waveforms, and even harmonics that are 'out-of-phase' tend to produce non-symmetrical waveforms. Finally, a perfect squarewave would have to contain an infinite number of odd harmonics, each of the appropriate amplitude. It is therefore impossible to produce a perfect squarewave, although a very close approximation can be achieved. It also has implications regarding the bandwidth required for a transmission system.

Bandwidth refers to the range of frequencies that a system can transmit with approximately equal attenuation or amplification. If the amplitudes of the harmonics are altered by differing amounts, then the complex wave shape will be changed. The received waveform will therefore be a distorted version of the original

The measurement of a complex waveform voltage using an analogue moving coil meter will normally result in a large error. This is because such an instrument is calibrated to indicate the rms value of a sinusoid, by employing a correction factor equal to the form factor for a sinewave. Depending upon the type of circuitry employed, electronic and digital instruments may introduce either zero error or a large error due to the complexity of the waveform. An instrument that is quoted as being a 'true rms' meter should introduce no error due to waveshape. However, in the case of this group of instruments, the bandwidth of the internal amplifier may introduce errors.

The best means of measuring a complex waveform is probably by the use of an oscilloscope. This gives the added advantage that the actual waveshape can also be observed. However, for the displayed waveform to be a faithful reproduction of the measured waveform *all* of the harmonics present in the original must be amplified (or attenuated) by the same amount. If this is not the case then the waveshape will be changed. Thus the amplifiers employed must have a very wide bandwidth, with a flat frequency response curve. The other main problem concerns any time delay (phase shift) that the circuitry imposes on different harmonics. A slight change of phase of a harmonic can result in a dramatic change in waveshape. For these reasons the amplifiers used in oscilloscopes are usually directly coupled. This avoids the use of coupling capacitors and resistors (which can introduce time delays and phase shift) and also allows the amplification of signals down to 0 Hz. Since most complex waveforms contain a d.c. component, it is important that this is treated in the same way as all the harmonics. In order to obtain an appreciation of the changes produced in a complex wave due to variations in the amplitude and phase of harmonics, it is suggested that experimentation with a waveform synthesiser and oscilloscope be undertaken.

7.8 Measurement of Phase and Frequency

These two characteristics of alternating quantities may be measured by meters designed specifically for this, i.e. a phasemeter and a frequency meter/counter. These instruments are not always readily available, and these measurements are more often performed with an oscilloscope, as will now be described.

The measurement of the phase relationship between two waveforms (of the same frequency) may be carried out with a CRO using either one of two methods.

Dual-trace method

The two waveforms are displayed on the screen, and by means of the 'Y'-shift controls the two traces are aligned along a common horizontal axis, as shown in Fig. 7.8. The time intervals T and t are then measured, using the graticule and timebase setting. Since the

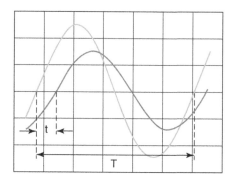

Fig. 7.8

periodic time T of a waveform corresponds to 360°, then the phase angle ϕ may be determined from:

$$\frac{\phi}{360} = \frac{t}{T}$$

Lissajous figures

In this method the CRO timebase is turned off, and one of the two waveforms is connected to the 'X' input. The second waveform is connected to the 'Y'-amplifier input as normal. Depending upon the phase angle between the two waveforms the resulting display will be either a straight line, a circle, or an ellipse. Examples are shown in Fig. 7.9, where the phase relationships are:

(a) $\phi = 0°$ (in phase)
(b) $\phi = 90°$
(c) $\phi = 180°$ (in antiphase)
(d) $0° < \phi < 90°$

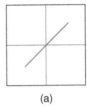 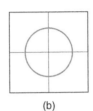 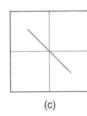 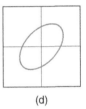

(a)   (b)   (c)   (d)

Fig. 7.9

Where the display is an ellipse, the phase angle may be calculated by measuring the dimensions A and B, as in Fig. 7.10, and applying the following:

$$\sin \phi = \frac{A}{B}$$

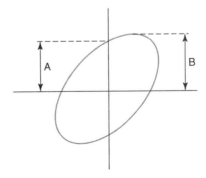

Fig. 7.10

7.9   A.C. Bridges

An a.c. bridge utilises the same basic principles as those of the Wheatstone bridge used for the d.c. measurement of resistance. The bridge supply will of course be an alternating supply, and the detector in the central limb must be sensitive to a.c. The balance condition occurs when the current through the detector is zero. The balance condition may also be expressed in terms of the ratio of the impedances of the four outer limbs, as below. The basic arrangement is shown in Fig. 7.11.

$$\frac{Z_1}{Z_2} = \frac{Z_3}{Z_4}$$

One of the limbs will contain the impedance to be measured (e.g. an inductor), and the other arms will contain variable reactance and resistance elements. Obtaining the balance condition is more difficult than for the simple Wheatstone bridge because the potentials at B and D must be equal in both amplitude and phase. There are a number of different forms of a.c. bridge for the measurement of inductors and capacitors, including the Owen, Hay's, Maxwell, and Schering bridges. Each of these may be used to measure either inductance or capacitance. More commonly, 'Universal' bridges that may be used to measure inductance, capacitance, resistance and Q-factor are employed. It is suggested that the student carries out measurements using a Universal bridge, since the practical experience thus gained will be far more instructive than a description of the procedure given here.

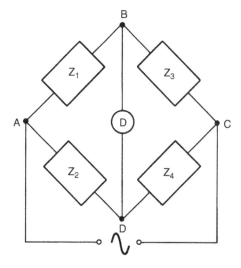

Fig. 7.11

Chapter 8

D.C. Transients

Learning Outcomes

This chapter explains the response of capacitor-resistor, and inductor-resistor circuits, when they are connected to and disconnected from, a d.c. supply. The affect of the time constant of *CR* circuits on a.c waveforms and the application to integrating and differentiating circuits is also explained.

On completion of the chapter, you should be able to:

1 Explain how the current and capacitor voltage in a series *C-R* circuit varies with time, when connected to/disconnected from a d.c. supply.
2 Explain how the current through, and p.d. across an inductor in a series *L-R* circuit varies with time, when connected to/disconnected from a d.c. supply.
3 Define the term time constant for both types of above circuits.
4 Carry out calculations in order to predict the value of current and/or voltage at any instant of time.
5 Carry out calculations in order to predict the time taken for the current and/or voltage to reach a specified value.
6 Demonstrate the effect of circuit time constant on rectangular waveforms, and relate these effects to the concepts of simple integration and differentiation.

8.1   Capacitor-Resistor Series Circuit (Charging)

Before dealing with the charging process for a *C-R* circuit, let us firstly consider an analogous situation. Imagine that you need to inflate a 'flat' tyre with a foot pump. Initially it is fairly easy to pump air into the tyre. However, as the air pressure inside the tyre builds up, it becomes progressively more difficult to force more air in. Also, as the internal pressure builds up, the rate at which air can be pumped in decreases. Comparing the two situations, the capacitor (which is to be charged) is analogous to the tyre; the d.c. supply behaves like the pump; the charging current compares to the air flow rate; and the p.d. developed between the plates of the capacitor has the same effect as the tyre pressure. From these comparisons we can conclude that as the capacitor voltage builds up, it reacts against the emf of the supply, so slowing down the charging rate. Thus, the capacitor will charge at a non-uniform rate, and will continue to charge until the p.d. between its plates is equal to the supply emf. This last point would also apply to tyre inflation, when the tyre pressure reaches the maximum pressure available from the pump. At this point the air flow into the tyre would cease. Similarly, when the capacitor has been fully charged, the charging current will cease.

Let us now consider the *C-R* charging circuit in more detail. Such a circuit is shown in Fig. 8.1. Let us assume that the capacitor is initially fully discharged, i.e. the p.d. between its plates (v_C) is zero, as will be the charge, *q*. Note that the lowercase letters *v* and *q* are used because, during the charging sequence, they will have continuously changing values, as will the p.d. across the resistor (v_R) and the charging current, *i*. Thus these quantities are said to have *transient* values.

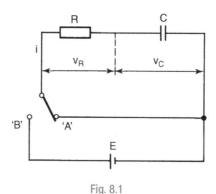

Fig. 8.1

At some time *t* = 0, let the switch be moved from position 'A' to position 'B'. At this instant the charging current will start to flow. Since there will be no opposition offered by capacitor p.d. ($v_C = 0$), then only the resistor, *R*, will offer any opposition. Consequently, the initial charging current (I_o) will have the maximum possible value for the circuit. This initial charging current is therefore given by:

$$I_o = \frac{E}{R} \text{ amp} \tag{8.1}$$

Since we are dealing with a series d.c. circuit, then the following equation must apply *at all times*:

$$E = v_R + v_C \text{ volt}\dots\dots\dots[1]$$

thus, at time *t* = 0

$$E = v_R + 0$$

i.e. the full emf of *E* volt is developed across the resistor at the instant the supply is connected to the circuit. Since $v_R = iR$, and at time *t* = 0, $i = I_o$, this confirms equation 8.1 above.

Let us now consider the situation when the capacitor has reached its fully-charged state. In this case, it will have a p.d. of E volt, a charge of Q coulomb, and the charging current, $i = 0$. If there is no current flow then the p.d. across the resistor, $v_R = 0$, and eqn [1] is:

$$E = 0 + v_C$$

Having confirmed the initial and final values for the transients, we now need to consider how they vary, with time, between these limits. It has already been stated that the variations will be non-linear (i.e. not a straight line graph). In fact the variations follow an *exponential* law. This fact can be proven by the following mathematical derivation. However, please note that, for a BTEC level 3 course, you are *not* required to reproduce this derivation.

$$E = v_R + v_C; \text{ where } v_R = Ri$$

$$\text{therefore, } E = Ri + v_C; \text{ but } i = \frac{dq}{dt}$$

$$\text{so, } E = R\frac{dq}{dt} + v_C$$

$$\text{but, } q = Cv_C; \text{ so } \frac{dq}{dt} = C\frac{dv_C}{dt}$$

$$\text{hence, } E = CR\frac{dv_C}{dt} + v_C$$

$$CR\frac{dv_C}{dt} = E - v_C$$

$$\text{so } dt = CR\left(\frac{1}{E - v_C}\right)dv_C$$

and integrating both sides of the above equation:

$$t = CR\int\left(\frac{1}{E - v_C}\right)dv_C$$

$$\text{so } t = -CR\ \ell n(E - v_C) + k\ldots\ldots\ldots[2]$$

where k is the constant of integration, which is determined as follows:

when $t = 0$, then $v_C = 0$, and substituting into [2] above:

$$0 = -CR\ \ell n\ E + k; \text{ hence, } k = CR\ \ell n\ E$$

Thus, eqn [2] becomes:

$$t = CR\ \ell n\ E - CR\ \ell n\ (E - v_C)$$

$$t = CR\ \ell n\left(\frac{E}{E - v_C}\right) = CR\ \ell n\left(\frac{1 - v_C}{E}\right)$$

therefore, $\dfrac{t}{CR} = \ell n \left(\dfrac{\ell - v_C}{E} \right)$; and antilogging both sides:

$$e^{-t/CR} = 1 - \frac{v_C}{E}$$

$$\frac{v_C}{E} = 1 - e^{-t/CR}$$

hence, $v_C = E(1 - e^{-t/CR})$ volt

From the above equation it may be seen that the capacitor voltage, v_C, changes exponentially with time. Also, since we know that the capacitor voltage increases from zero to E volt, then during the transient period it follows an exponential growth.

The term CR in the exponent for e is known as the time constant, τ (Greek letter tau), of the system. It may appear strange that the product of capacitance and resistance yields a result having units of time. This may be justified by considering a simple dimensional analysis, as follows.

$$C = \frac{Q}{V} = \frac{It}{V} \text{ and } R = \frac{V}{I}$$

$$\text{so, } CR = \frac{It}{V} \times \frac{V}{I} = t \text{ seconds}$$

The time constant of the capacitor-resistor arrangement may be defined as the time that it would take the capacitor to become fully charged, *if* it continued to charge at the *initial* rate.

Since $q = Cv_C$, then it follows that the charge on the capacitor also grows exponentially, from zero to Q coulomb. Thus, the equations for capacitor voltage and charge are both of the form:

$$v_C = E(1 - e^{-t/\tau}) \text{ volt} \tag{8.2}$$

$$\text{and } q = Q(1 - e^{-t/\tau}) \text{ coulomb} \tag{8.3}$$

As the applied emf, E, must at all times be equal to the sum of the p.d.s across the capacitor and the resistor, then it follows that the resistor voltage must *decay* exponentially, from E volt to zero. The equation for this p.d. is:

$$v_R = E e^{-t/\tau} \text{ volt} \tag{8.4}$$

Similarly, since the p.d. across the resistor is directly proportional to the current flowing through it, then the charging current must decay exponentially from I_o to zero. The equation for the circuit current is therefore:

$$i = I_o \, e^{-t/\tau} \text{ amp} \tag{8.5}$$

The graphs showing how the above four quantities vary, after the supply is connected, are Figs. 8.2 to 8.5 inclusive. The following points of interest arise from these graphs.

(i) For exponential growth, the quantity concerned reaches 63.2% of its final value after the elapse of one time constant.

(ii) For exponential decay, the quantity concerned falls to 36.8% of its initial value after the elapse of one time constant.

(iii) After five time constants ($t = 5\tau$ seconds), the quantity concerned has reached a value within 0.67% of its final value, e.g. $v_C = 0.993 \, E$ volt.

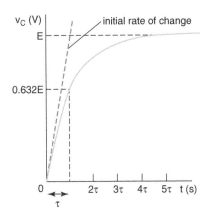

Fig. 8.2

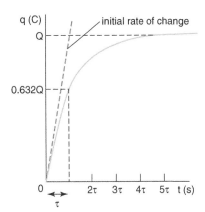

Fig. 8.3

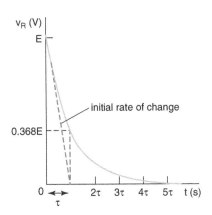

Fig. 8.4

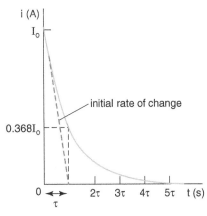

Fig. 8.5

Theoretically, it is a feature of an exponential function that the capacitor voltage can never actually achieve a value of E volt. However, for practical purposes, it is assumed that the capacitor reaches its fully charged state in a time of 5τ seconds.

(iv) Considering *any* point on the graph, it would take one time constant for that quantity to reach its final value *if* it continued to change at the same rate as at that point. Thus the exponential graph may be considered as being formed from an infinite number of straight lines, each of which represents the slope at a particular instant of time. This is illustrated in Fig. 8.6.

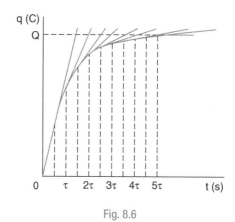

Fig. 8.6

Points (i) to (iii) above may be confirmed from the relevant equations, as follows:

When $t = \tau$ seconds (after the elapse of one time constant), for an exponential growth:

$$v_C = E(1 - e^{-1}) = E(1 - 0.368) = 0.632\,E$$

and for an exponential decay:

$$i = I_o e^{-1} = 0.368\,I_o$$

and when $t = 5\tau$ seconds:

$$v_C = E(1 - e^{-5}) = E(1 - 6.74 \times 10^{-3}) = 0.993\,E$$

Worked Example 8.1

Q   An $8\,\mu$F capacitor is connected in series with a $0.5\,\text{M}\Omega$ resistor, across a 200 V d.c. supply. Calculate (a) the circuit time constant, (b) the initial charging current, (c) the p.d.s across the capacitor and resistor 6 seconds after the supply is connected, and (d) the time taken for the capacitor p.d. to reach 160 V. You may assume that the capacitor is initially fully discharged.

A

$$C = 8 \times 10^{-6}\,\text{F}; R = 0.5 \times 10^{6}\,\Omega; E = 200\,\text{V}$$

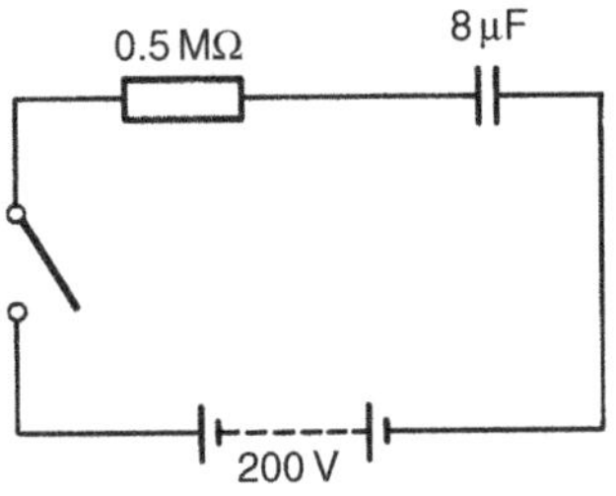

Fig. 8.7

(a) $\tau = CR$ second $= 8 \times 10^{-6} \times 0.5 \times 10^{6}$

so $\tau = 4$ s **Ans**

(b) $I_o = \dfrac{E}{R}$ amp $= \dfrac{200}{0.5 \times 10^{6}}$

therefore $I_o = 400$ µA **Ans**

(c) $v_C = E(1 - e^{-t/\tau})$ volt $= 200\,(1 - e^{-6/4})$

$v_C = 155.37$ V **Ans**

$v_R = Ee^{-t/\tau}$ volt $= 200e^{-6/4}$

so $v_R = 44.63$ V **Ans**

Or, more simply, $v_R = E - v_C$ volt

$= 200 - 155.37$

so $v_R = 44.63$ V as before

(d) $v_C = E(1 - e^{-t/\tau})$ volt

and to find the time t, first transpose the equation to make the term $e^{-t/\tau}$ the subject:

$$\frac{v_C}{E} = 1 - e^{-t/\tau}$$

$$\text{so, } e^{-t/\tau} = 1 - \frac{v_C}{E}$$

$$e^{-t/4} = 1 - \frac{160}{200} = 0.2$$

and taking logs (to the base e) of both sides:

$$-\frac{t}{4} = \ell n\ 0.2 = -1.609$$

hence $t = 4 \times 1.609 = 6.44$ s **Ans**

Note: It may have been slightly simpler to use the fact that when $v_C = 160$ V, then $v_R = 40$ V, and then use the equation

$$v_R = Ee^{-t/\tau}$$

It is left to the reader to carry out this calculation to verify the answer to part (d) above.

8.2   Capacitor-Resistor Series Circuit (Discharging)

Consider the circuit of Fig. 8.1, where the switch has been in position 'B' for sufficient time to allow the charging process to be completed. Thus the charging current will be zero, the p.d. across the resistor will be zero, the p.d. across the capacitor will be E volt, and it will have stored a charge of Q coulomb.

At some time $t = 0$, let the switch be moved back to position 'A'. The capacitor will now be able to discharge through resistor R. The general equation for the voltages in the circuit will still apply.

In other words $E = v_R + v_C$

but, at the instant the switch is moved to position 'A', the source of emf is removed. Applying this condition to the general equation above yields:

$$0 = v_R + v_C; \quad \text{where } v_C = E \text{ and } v_R = I_o R$$
$$\text{so } 0 = I_o R + E$$

$$\text{hence } I_o = -\frac{E}{R} \text{ amp} \tag{8.6}$$

This means that the initial discharge current has the same value as the initial charging current, but (as you would expect) it flows in the opposite direction.

Since the capacitor is discharging, then its voltage will decay from E volt to zero; its charge will decay from Q coulomb to zero; and the discharge current will also decay from I_o to zero. The equations for all the quantities will be as follows:

$$v_C = Ee^{-t/\tau} \text{ volt} \tag{8.7}$$

$$q = Qe^{-t/\tau} \text{ coulomb} \tag{8.8}$$

$$i = -I_o e^{-t/\tau} \text{ amp} \tag{8.9}$$

$$v_R = Ee^{-t/\tau} \text{ volt} \tag{8.10}$$

The graphs for v_C and i are shown in Fig. 8.8.

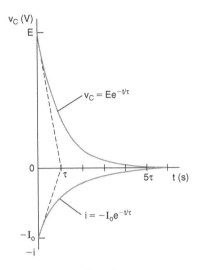

Fig. 8.8

Note: The time constant for the *C-R* circuit was defined previously in terms of the capacitor charging. However, a time constant also applies to the discharge conditions. It is therefore better to define the time constant in a more general manner, as follows:

> The time constant of a circuit is the time that it would have taken for any transient variable to change, from one steady state to a new steady state, if it had maintained its original rate of change.

Worked Example 8.2

Q A *C-R* charge/discharge circuit is shown in Fig. 8.9. The switch has been in position '1' for a sufficient time to allow the capacitor to become fully discharged. (a) If the switch is now moved to position '2', calculate the capacitor p.d. after 0.3 s. (b) If after the 0.3 s the switch is now returned to position '1', calculate the value of discharge current and capacitor p.d. after a further period of 0.3 s.

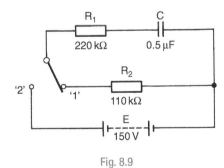

Fig. 8.9

A

$C = 0.5\,\mu F;\ R_1 = 220\,k\Omega;\ R_2 = 110\,k\Omega;\ E = 150V$

(a) When charging, only resistor R_1 is connected in series with the capacitor, so R_2 may be ignored.

$$\tau = CR_1 \text{ seconds} = 0.5 \times 10^{-6} \times 220 \times 10^3$$

$$\text{so } \tau = 0.11 \text{ s}$$

$$v_C = E(1 - e^{-t/\tau}) \text{ volt}$$

$$= 150(1 - e^{-0.3/0.11})$$

$$\text{hence } v_C = 140.2 \text{ V } \textbf{Ans}$$

(b) When discharging, both R_1 and R_2 are connected in series with the capacitor, so their combined resistance $R = R_1 + R_2$, will determine the discharge time constant.

$$\tau = CR \text{ seconds} = 0.5 \times 10^{-6} \times 330 \times 10^3$$

$$\text{so, } \tau = 0.16 \text{ s}$$

Also, the capacitor will be discharging from a voltage of 140.2 V. Thus, the initial discharge current will be:

$$I_o = \frac{140.2}{330 \times 10^3} \text{ amp}$$

$$= 424.85 \text{ μA}$$

Therefore after 0.3 s of the discharge cycle, the current will be:

$$i = I_o\, e^{-t/\tau} \text{ amp} = 424.85\, e^{-0.3/0.16} \text{ μA}$$

$$\text{hence, } i = 65.15 \text{ μA } \textbf{Ans}$$

$$v_C = 140.2\, e^{-0.3/0.16} \text{ volt}$$

$$\text{so } v_C = 21.5 \text{ V } \textbf{Ans}$$

Note: It should be obvious from equations (8.7) and (8.10) that, in the discharge circuit, the p.d. across the capacitor and the circuit resistance must be the same value. Thus the last part of (b) could have been obtained thus:

$$v_C = v_R = iR = 65.15 \times 10^{-6} \times 330 \times 10^3$$

$$\text{hence, } v_C = 21.5 \text{ V as above.}$$

Worked Example 8.3

Q   Five 10 μF capacitors are connected in parallel and charged to a p.d. of 150 V. This capacitor bank is to be discharged by connecting a resistor across it, such that the discharge process is completed in approximately 15 seconds. Calculate (a) the resistor value required, (b) the initial value of the discharge current, and (c) the current flowing 12 s after connecting the resistor.

A

$$C_1 = C_2 = C_3 = C_4 = C_5 = 10 \times 10^{-6}\, \text{F}; V = 150\text{V}; t_1 = 15 \text{ s}; t_2 = 12 \text{ s}$$

(a) Since the capacitors are connected in parallel the total capacitance of the bank, C, is simply the sum of them.

$$C = 5 \times 10\mu F = 50\,\mu F$$

For practical purposes, the discharge process takes approximately five time constants, so

$$5\tau = t_1 \text{ seconds}$$

$$\text{and } \tau = \frac{t_1}{5} = \frac{15}{3} = 3\text{ s}$$

$$\text{also, } \tau = CR \text{ seconds}$$

$$\text{so, } R = \frac{\tau}{C}\text{ohm} = \frac{3}{5 \times 10^{-5}}$$

$$R = 60 \text{ k}\Omega \text{ Ans}$$

(b) $\quad I_o = \dfrac{V}{R} \text{ amp} = \dfrac{150}{60 \times 10^3}$

$I_o = 2.5$ mA **Ans**

(c) $\quad i_2 = I_o\, e^{-t_2/\tau} \text{ amp} = 2.5\, e^{-12/3} \text{ mA}$

$\quad\quad = 2.5\, e^{-4}$

$i_2 = 45.8 \ \mu A$ **Ans**

8.3   Inductor-Resistor Series Circuit (Connection to Supply)

Consider the circuit of Fig. 8.10. At some time $t = 0$, the switch is moved from position 'A' to position 'B'. The connection to the supply is now complete, and current will start to flow, increasing towards its final steady value. However, whilst the current is *changing* it will induce a

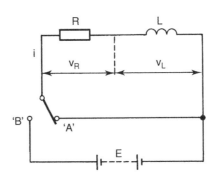

Fig. 8.10

back-emf across the inductor, of e volt. From electromagnetic induction theory we know that this induced emf will have a value given by:

$$e = -L\frac{\mathrm{d}i}{\mathrm{d}t} \text{ volt}$$

Being a simple series circuit, Kirchhoff's voltage law will apply, such that the sum of the p.d.s equals the applied emf. Also, since we are considering a perfect inductor (the resistor shown may be considered as the coil's resistance), the p.d. across the inductor will be exactly equal but opposite in polarity to the induced emf.

$$\text{Therefore, } v_L = -e = L\frac{\mathrm{d}i}{\mathrm{d}t} \text{ volt}$$

$$\text{hence, } E = v_R + v_L \text{ volt}$$

$$\text{or, } E = iR + L\frac{\mathrm{d}i}{\mathrm{d}t} \text{ volt.........[1]}$$

Comparing this equation with that for the *C-R* circuit, it may be seen that they are both of the same form. Both contain a first-derivative term (a 'd/dt' term) and a 'constant' term. Using the analogy technique, we can conclude that both systems will respond in a similar manner. In the case of the *L-R* circuit, the *current* will increase from zero to its final steady value, following an exponential law.

At the instant that the switch is moved from 'A' to 'B' ($t = 0$), the current will have an instantaneous value of zero, but it *will* have a certain rate of change, di/dt amp/s. From eqn [1] above, this initial rate of change can be obtained, thus:

$$E = 0 + L\frac{\mathrm{d}i}{\mathrm{d}t}$$

$$\text{so, initial } \frac{\mathrm{d}i}{\mathrm{d}t} = \frac{E}{L} \text{ amp/s} \qquad (8.11)$$

When the current reaches its final steady value, there will be no back-emf across the inductor, and hence no p.d. across it. Thus the only limiting factor on the current will then be the resistance of the circuit. The final steady current is therefore given by:

$$I = \frac{E}{R} \text{ amp} \qquad (8.12)$$

The time constant of the circuit is obtained by dividing the inductance by the resistance.

$$\text{Thus } \tau = \frac{L}{R} \text{ seconds} \qquad (8.13)$$

The above equation may be confirmed by using a simple form of dimensional analysis, as follows.

$$\text{In general, } V = \frac{LI}{t}; \text{ so } L = \frac{Vt}{I}$$

$$\text{and } R = \frac{V}{I}$$

$$\text{therefore, } \frac{L}{R} = \frac{Vt}{I} \times \frac{I}{V} = t \text{ seconds}$$

The time constant of the circuit may be defined in the general terms given in the '*Note*', in the previous section, dealing with the *C-R* circuit.

The rate of change of current will be at its maximum value at time $t = 0$, so the p.d. across the inductor will be at its maximum value at this time. This p.d. therefore decays exponentially from E volt to zero. The graphs for i, v_R, and v_L are shown in Figs. 8.11 to 8.13, respectively.

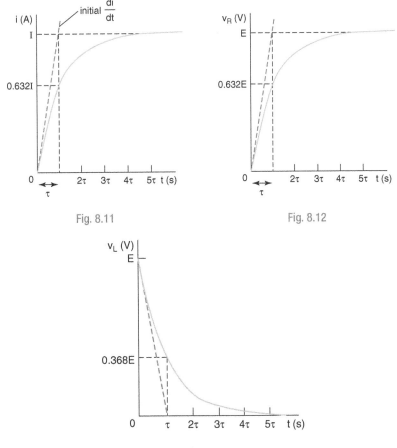

Fig. 8.11

Fig. 8.12

Fig. 8.13

Worked Example 8.4

Q   The field winding of a 110V, d.c. motor has an inductance of 1.5H, and a resistance of 220Ω. From the instant that the machine is connected to a 110V supply, calculate (a) the initial rate of change of current, (b) the final steady current, (c) the current 10ms after connection, and (d) the time taken for the current to reach its final steady value.

A

$E = 110V; L = 1.5\ H; R = 220\ \Omega; t = 10\ ms$

The circuit diagram is shown in Fig. 8.14.

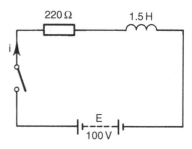

Fig. 8.14

(a)   initial $\dfrac{di}{dt} = \dfrac{E}{L}$ amp/s $= \dfrac{110}{1.5}$

so, initial $\dfrac{di}{dt} = 73.33$ A/s **Ans**

(b)   final current, $I = \dfrac{E}{R}$ amp $= \dfrac{110}{220}$

therefore, $I = 0.5$ A **Ans**

(c) $\tau = \dfrac{L}{R}$ second $= \dfrac{1.5}{220}$

hence, $\tau = 6.82$ ms

$i = I(1 - e^{-t/\tau})$ amp

$= 0.5(1 - e^{-10/6.82})$

therefore, $i = 0.385$ A **Ans**

(d)   Since the system takes approximately 5τ seconds to reach its new steady state, then the current will reach its final steady value in a time:

$t = 5 \times 6.82$ ms $= 34.1$ ms **Ans**

8.4   Inductor-Resistor Series Circuit (Disconnection)

Figure 8.15 shows such a circuit, connected to a d.c. supply. Assume that the current has reached its final steady value of I amps. Let the switch now be returned to position 'A' (at time $t = 0$). The current will now decay to zero in an exponential manner. However, the decaying current will induce a back-emf across the coil. This emf must oppose the change of current. Therefore, the decaying current will flow *in the same direction* as the original steady current. In other words, the back-emf will try to maintain the original current flow. The graph of the decaying current, with respect to time, will therefore be as shown in Fig. 8.16. The time constant of the circuit will, of course, still be L/R second, and the current will decay from a value of $I = E/R$ amp. The initial rate of decay will also be E/L amp/s.

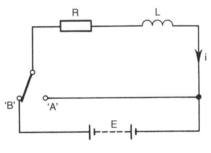

Fig. 8.15

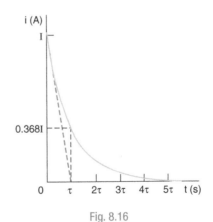

Fig. 8.16

Worked Example 8.5

Q   A coil of resistance $125\,\Omega$ and inductance L henry is connected to a 200 V d.c. supply, and the resulting initial rate of change of current is 800 A/s. Calculate (a) the coil inductance, (b) the coil time constant, (c) the final steady current, and (d) the time taken for the current to reach 1.4 A.

A

$$R = 125\,\Omega; V = 200\text{V}; \frac{di}{dt} = 800 \text{ A/s}; i = 1.4 \text{ A}$$

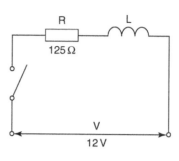

Fig. 8.17

(a) Initial $\dfrac{di}{dt} = \dfrac{V}{L}$ amp/second

so $800 = \dfrac{200}{L}$

and $L = \dfrac{200}{800}$

$L = 0.25$ H **Ans**

(b) $\tau = \dfrac{L}{R}$ second $= \dfrac{0.25}{125} = 2$ ms **Ans**

(c) The final *steady* current is simply limited by the resistance of the coil, so

$I = \dfrac{V}{R}$ amp $= \dfrac{200}{125}$

$I = 1.6$ A **Ans**

(d) $i = I\,(1 - e^{-t/\tau})$ amp

$1.4 = 1.6(1 - e^{-t/0.002})$

$1.4 = 1.6(1 - e^{-500t})$

$0.875 = 1 - e^{-500t}$

$e^{-500t} = 1 - 0.875 = 0.125$

and taking logs of both sides:

$-500t = \ell n\, 0.125 = -2.0794$

$t = \dfrac{2.0794}{500}$

$t = 4.16$ ms **Ans**

Worked Example 8.6

Q   Figure 8.18 represents a relay coil that is energised/de-energised by means of a 2-pole switch, which has been in the position shown for some considerable time. In order for the relay to close the contact, the coil current must be 0.8 A. When de-energised, the contact opens when the coil current has fallen to 0.5 A. Calculate (a) the time taken for the contact to close, and (b) the time taken for the contact to open, if the switch is left in position '2' for 25 ms, before it is returned to position '1'. Sketch a graph showing the variation of current, versus time, for a total of 35 ms after the supply is first connected, showing all principal values.

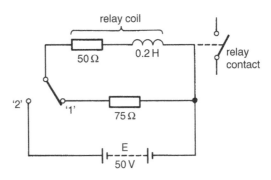

Fig. 8.18

A

$L = 0.2$ H; $R_1 = 50\,\Omega$; $R_2 = 125\,\Omega$; $E = 50$V; $i_1 = 0.8$ A; $i_2 = 0.5$ A

(a)   For current growth, $\tau = \dfrac{L}{R_1}$ second $= \dfrac{0.2}{50}$

so, $\tau = 4$ ms

Final steady current, $I = \dfrac{E}{R_1}$ amp $= \dfrac{50}{50} = 1$ A

Now, $i_1 = I(1 - e^{-t/\tau})$ amp

$$\frac{i_1}{I} = 1 - e^{-t/\tau}$$

$$e^{-t/\tau} = 1 - \frac{i_1}{I}$$

$$\frac{-t}{\tau} = \ell n\left(1 - \frac{i_1}{I}\right)$$

therefore, $-t = \tau\, \ell n\left(1 - \frac{i_1}{I}\right)$ second

$$= 4\, \ell n(1 - 0.8)$$

hence, $t = 6.438$ ms **Ans**

(b)   The current will grow to its final value in $5\tau = 20$ ms. Since the switch is left in position '2' for 25 ms, then the current will be at its steady value of 1A when the switch is returned to position '1'. Hence, the initial decay current will be $I = 1$ A. However, in position '1', the decay current will flow through both the 50 Ω and the 75 Ω resistor. That is to say, the total resistance, $R_2 = 125\,\Omega$. This means that the decay current time constant is given by:

$$\tau = \frac{L}{R_2}\ \text{second} = \frac{0.2}{125} = 1.6\ \text{ms}$$

Also, $i_2 = Ie^{-t/\tau}$ amp

$$\frac{i_2}{I} = e^{-t/\tau}$$

$$-t = \tau\ \ell n\frac{i_2}{I} = 1.6\ \ell n\ 0.5\ \text{ms}$$

therefore, $t = 1.11\ \text{ms}$ **Ans**

The current decays to zero in approximately 8 ms, and the graph of current versus time will be as shown in Fig. 8.19.

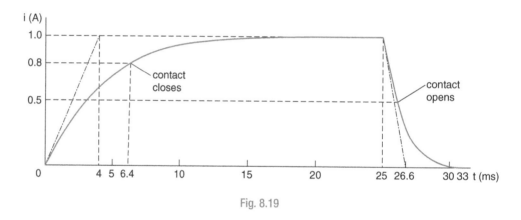

Fig. 8.19

8.5   Affect of Time Constant on Alternating Waveforms

Consider a series $C\text{-}R$ circuit, with a rectangular waveform applied to the input terminals, as in Fig. 8.20. The p.d.s developed across the resistor and capacitor are monitored on a double-beam oscilloscope. If the resistor is adjusted to a low value, then the circuit time constant ($\tau = CR$) will be short. Let us assume that, in this case, τ is very much less than the half-period ($T/2$) of the input waveform. Since the time constant is very short, then the capacitor can charge and discharge rapidly. The waveform for the p.d. developed across C is shown in

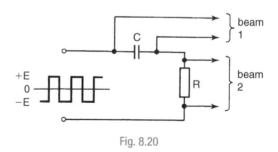

Fig. 8.20

Fig. 8.21. Also shown is the p.d. developed across R. Notice that if these two waveforms are added, the input rectangular waveform is reproduced. This confirms the fact that $E = v_R + v_C$ volt.

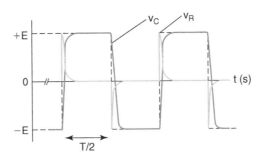

Fig. 8.21

Let the time constant now be increased, by increasing the value of R. Let us also assume that the new time constant is sufficiently long so as to prevent the capacitor reaching its fully charged state of E volt. The effect on the p.d.s is illustrated in Fig. 8.22.

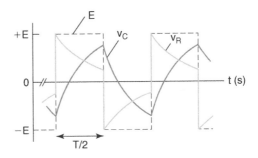

Fig. 8.22

Finally, let the time constant be further increased, to the point where τ is very much greater than $T/2$. The resulting waveforms are shown in Fig. 8.23. In this situation the capacitor charges and discharges so slowly that the waveform of v_C shows only the initial rates of change of v_C, which are a close approximation to straight lines.

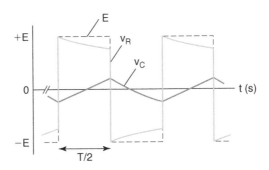

Fig. 8.23

8.6   Differentiating and Integrating Networks

Consider the network of Fig. 8.24. The input waveform is a rectangular wave, and the network output is taken across the resistor. Let the circuit time constant be very short compared with the half-period of the input. To meet this criterion, the 'rule of thumb' is that τ is less than, or equal to one tenth of the input half-period, i.e. $\tau \leq T/20$. The resulting output waveform is shown in Fig. 8.25.

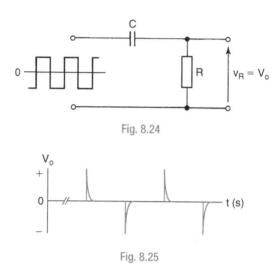

Fig. 8.24

Fig. 8.25

Now, the mathematical process of differentiation is merely a means of determining the rate of change of a quantity. For example, velocity is the rate of change of distance with time. Expressed mathematically this is written as $v = ds/dt$.

A *perfect* rectangular waveform has vertical rising and falling edges, and horizontal 'tops'. However, vertical edges implies instantaneous changes from one value to another, i.e. a change in zero time. Such a change is a physical impossibility, since the rate of change would be *infinite*. In practice, the rising and falling edges of a rectangular waveform are very steep, but cannot be absolutely vertical. The 'tops' of the waveform can be horizontal, since this is a zero rate of change. However, if a perfect rectangular waveform was differentiated, the result would be a series of vertical 'spikes' of infinite length and of zero width (occupying zero time). This is illustrated in Fig. 8.26. If we

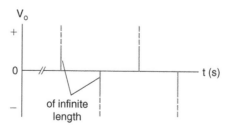

Fig. 8.26

now compare this waveform with that of Fig. 8.25, it may be seen that the output voltage of the network is a very close approximation to the differential of the input waveform. Indeed, the shorter the circuit time constant, the 'better' the differentiating effect.

This simple form of differentiating network is used in some electronic circuits to convert a relatively wide rectangular pulse into a 'spike'. This technique is often used to generate the trigger pulses to electronic timing circuits, such as a device known as a 555 timer.

Let us now transpose the places of C and R in the circuit, and also alter their values to provide a very long time constant. In this case, the 'rule of thumb' is that τ is equal to, or greater than, ten times the half-period of the input waveform, i.e. $\tau \geq 5T$ seconds. The circuit arrangement is shown in Fig. 8.27, and the resulting output waveform is shown in Fig. 8.28. If a rectangular waveform is integrated, then a triangular waveform, of reduced amplitude, results. Thus this circuit is an integrating network. This reduction of amplitude may be confirmed by considering the integration of a sinewave, as follows.

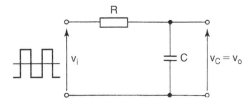

Fig. 8.27

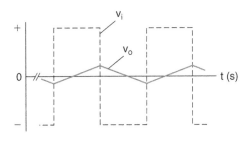

Fig. 8.28

Let the input, $v_i = V_{im} \sin \omega t$ volt

and the output, $v_o = \int v_i \, dt$

then, $v_o = \int V_{im} \sin \omega t \, dt$ volt

hence, $v_o = -\dfrac{V_{im}}{\omega} \cos \omega t$ volt

i.e. the amplitude is *reduced* by a factor of ω.

The integrating action of the circuit may be more easily understood by considering a sinusoidal input, as in Fig. 8.29. In order to achieve a long time constant we could have large values for both the resistor and the capacitor. Since X_C is proportional to $1/C$, then for a large value capacitor, X_C will be very much less than R. The resulting phasor diagram is shown in Fig. 8.30. From this diagram it may be seen that $V \approx V_R$, and the 'output', V_C, lags V by a large angle θ, which is approaching $90°$. The corresponding waveform diagram is shown in Fig. 8.31. From this it may be seen that the output, V_C, has a very small amplitude, and is a negative cosine wave. This is precisely the mathematical result shown above, which confirms that the circuit acts as an integrator.

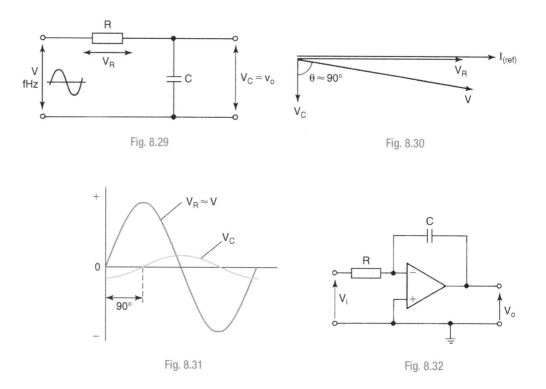

Fig. 8.29

Fig. 8.30

Fig. 8.31

Fig. 8.32

It is left to the reader to confirm the differentiating action for a sinusoidal input, using the appropriate phasor and waveform diagram. Remember though, for 'good' differentiation, a very short time constant is required.

In most practical applications using an integrator the severe reduction in amplitude of the output waveform is undesirabl. To overcome this problem an active integrator is used, which involves the use of an operational amplifier. A circuit of such an arrangement is shown in Fig. 8.32. This form of active integrator is utilised in applications such as analogue computers (simulators), some types of digital voltmeter, and oscilloscope timebase circuits. The latter two applications require the production of a linearly changing, or ramp, voltage.

It should be noted that series *L-R* circuits will also exhibit the differentiating and integrating properties described above. However, due to the relative weight, bulk and cost of large value inductors, integrating and differentiating networks using inductors are not deliberately used to achieve these effects. Any circuit containing inductors can of course cause the attenuation and distortion to rectangular waveforms described above, but this is usually an unwanted effect.

Worked Example 8.7

Q   A simple integrating network consists of a *C-R* circuit as shown in Fig. 8.33. The input waveform is a 2 kHz squarewave of amplitude 5 V.

(a)   By means of suitable calculations, verify that this circuit will provide 'good' integration.

(b)   Calculate the amplitude of the output waveform.

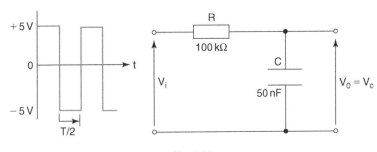

Fig. 8.33

A

$C = 50 \times 10^{-9}\,F; R = 10^5\,\Omega$

(a)   For 'good' integration the capacitor charge/discharge p.d. must be limited to the initial (approximately straight portions) of the charge/discharge curves. For this to occur the circuit time constant needs to be much greater than (at least ten times) the half-period, *T/2*, of the input waveform.

$$T = \frac{1}{f} \text{ second} = \frac{1}{2000} = 0.5 \text{ ms}$$

$$\text{so, } T/2 = 0.25 \text{ ms}$$

$$\tau = CR \text{ second} = 50 \times 10^{-9} \times 10^5$$

$$\tau = 5 \text{ ms}$$

$$\text{the ratio } \frac{\tau}{T/2} = \frac{5}{0.25} = 20$$

Thus the time constant of the circuit is 20 times that of the half-period of the input waveform, hence 'good' integration will result **Ans**

(b) Considering Fig. 8.34 it may be seen that the charge/discharge voltage for the capacitor varies between $+5\,V$ and $-5\,V$, i.e. a total change of $10\,V$ peak to peak. Thus the capacitor p.d. is 'aiming' to change by this amount, and would indeed do so if it had sufficient time (5τ seconds $= 25\,ms$) in which to do so. However, it has only $T/2$ seconds ($0.25\,ms$) allowed, so the total change in p.d. will be that achieved in $0.25\,ms$.

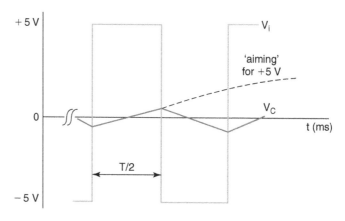

Fig. 8.34

$$v_{p-p} = V_{p-p}(1 - e^{-t/\tau}) \text{ volt } \text{ where } t = 0.25 \text{ ms}$$

$$= 10\,(1 - e^{-0.25/5}) = 10\,(1 - e^{-0.05})$$

$$= 10\,(1 - 0.9512)$$

$$v_{p-p} = 0.488 \text{ V}$$

$$\text{so, } v_p = 0.244 \text{ V } \textbf{Ans}$$

Summary of Equations

C-R circuits:

Time constant, $\tau = CR$ second

Initial current, $I_o = \dfrac{E}{R}$ amp

steady-state conditions after approx. 5τ second

For charging capacitor:

capacitor p.d. at any time t is, $v_C = E(1 - e^{-t/\tau})$ volt

circuit current at time t is, $i = I_o e^{-t/\tau}$ amp

after τ second, $v_C = 0.632E$ volt; and $i = 0.368I_o$ amp

For discharging capacitor:

$$v_C = Ee^{-t/\tau} \text{ volt}$$
$$i = -I_o e^{-t/\tau} \text{ amp}$$

L-R circuit:

$$\text{Time constant, } \tau = \frac{L}{R} \text{ second}$$

$$\text{Initial rate of change of current, } \frac{\mathrm{d}i}{\mathrm{d}t} = \frac{E}{L} \text{ amp/second}$$

$$\text{final current flowing, } I = \frac{E}{R} \text{ amp}$$

steady-state conditions after approx. 5τ second

On connection:

p.d. across inductor at time t is, $v_L = E\,e^{-t/\tau}$ volt

circuit current at time t is, $i = I(1 - e^{-t/\tau})$ amp

after τ second, $V_L = 0.368E$ volt; and $i = 0.632$ amp

On disconnection:

$$v_L = Ee^{-t/\tau} \text{ volt}$$
$$i = Ie^{-t/\tau} \text{ amp}$$

Assignment Questions

1 A 47 µF capacitor is connected in series with a 39 kΩ resistor, across a 24 V d.c. supply. Calculate (a) the circuit time constant, (b) the values for initial and final charging current, and (c) the time taken for the capacitor to become fully charged.

2 A 150 mH inductor of resistance 50 Ω is connected to a 50 V d.c. supply. Determine (a) the initial rate of change of current, (b) the final steady current, and (c) the time taken for the current to change from zero to its final steady value.

3 An inductor of negligible resistance and inductance 0.25 H, is connected in series with a 1.5 kΩ resistor, across a 24 V d.c. supply. Calculate (a) the current flowing after one time constant, (b) the p.d. across the inductor after two time constants, and (c) the p.d. across the resistor after three time constants.

4 An uncharged capacitor of 200 nF is connected to a 150 V d.c. supply via a 75 kΩ resistor. Determine the capacitor p.d. and circuit current 10 ms after connection.

5 An 8 µF capacitor is charged to 100 V, and then discharged through a 1.2 MΩ resistor. Calculate (a) the capacitor p.d. 2.5 s after discharging has commenced, and (b) the time taken for the voltage to fall to 20 V.

6 A 5 H inductor has a resistance R ohm. This inductor is connected in series with a 10 Ω resistor, across a 140 V d.c. supply. If the resulting circuit time constant is 0.4 s, determine (a) the value of the coil resistance, and (b) the current flowing 650 ms after connection to the supply.

7 Define the *time constant* of a capacitor-resistor series circuit.

   Such a circuit comprises a 50 µF capacitor and a resistor, connected to a 100 V d.c. supply via a switch. If the circuit time constant is to be 5 s, determine (a) the resistor value, (b) the initial charging current, (c) the time taken for the capacitor p.d. to reach 85 V, and (d) the value of the charging current 12 s after closing the switch.

8 The dielectric of a 20 µF capacitor has a resistance of 65 MΩ. This capacitor is fully charged from a 120 V d.c. supply. Calculate the time taken, after disconnection from the supply, for (a) the capacitor stored p.d. to fall to 99.8 V, and (b) the capacitor to become fully discharged.

9 A circuit consists of a 270 Ω resistor in parallel with a 4 H inductor, of resistance 30 Ω. If this combination is connected across a 100 V d.c. supply for 10 ms, and then disconnected, calculate the current flowing 5 ms after disconnection.

10 An electronic device is required to operate 1.5 s after a switch is closed. The device will operate when its input voltage (V_i) is equal to or greater than 5 V. This time delay is to be achieved by charging a capacitor via a resistor, as shown in Fig. 8.35. Determine the value to which the resistor must be adjusted in order to achieve the required time delay.

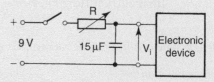

Fig. 8.35

11 A relay coil has an inductance of 120 mH and resistance 20 Ω. When it is connected to a 50 V d.c. supply, determine (a) the initial rate of change of current, (b) the time taken for the currernt to reach 1.5 A, and (c) the value of the current 4 ms after connection.

12 For the circuit shown in Fig. 8.36, determine the p.d. between terminals X and Y, (a) before the switch is closed, (b) at the instant that the switch is closed, and (c) 0.85 s after the switch is closed.

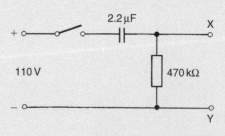

Fig. 8.36

13 The field winding of a 110 V d.c. motor has a resistance of 12 Ω, and a time constant of 2.5 s. Calculate (a) the inductance of the winding, (b) the time for the current to reach 1.8 A, and (c) the current flowing 1 s after connection to the 110 V supply.

14 For the circuit arrangement of Fig. 8.37, the capacitor is initially uncharged. At some time $t = 0$, the switch is closed for a period of 30 s;

Assignment Questions

after which it is re-opened. (a) Sketch (to scale) the variation of capacitor current, from $t = 0$ until the capacitor is again fully discharged. Show all principal values on your graph. (b) Calculate the charge on the capacitor 10 s after connection, and (c) the p.d. across the $750\,\Omega$ resistor 20 s after disconnection.

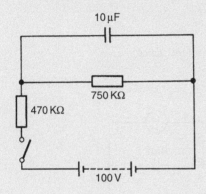

Fig. 8.37

15 A squarewave of frequency 1 kHz and amplitude 10 V is applied to terminals A and B of the network shown in Fig. 8.38. (a) On the same axes, sketch the graphs of V_{AB} and V_{CD}, (b) state the mathematical function carried out by this network, and (c) indicate the effect on V_{CD} if the resistor value is reduced to $2\,k\Omega$.

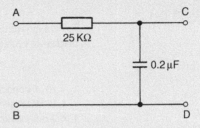

Fig. 8.38

16 A squarewave of frequency 1 kHz and amplitude 5 V is applied to terminals A and B of the network shown in Fig. 8.39. (a) Determine the maximum allowable value for the resistor, in order for the circuit to act as a 'good' differentiator, and (b) sketch, on the same axes, the input and output waveforms under this condition.

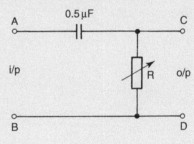

Fig. 8.39

Suggested Practical Assignments

Assignment 1

To investigate the variation of capacitor voltage and current during charge and discharge cycles.

Apparatus:

1 × 10 μF capacitor
1 × 10 MΩ resistor
1 × 2-pole switch
1 × d.c. power supply
1 × ammeter (microammeter)
1 × DVM (with highest possible input resistance)
1 × stopwatch

Method:

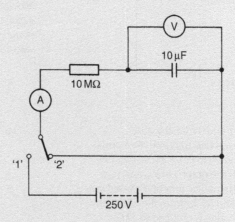

Fig. 8.40

1   Connect the circuit of Fig. 8.40, and adjust the psu output to 250V.
2   Simultaneously move the switch to position '1' and start the stopwatch.
3   Record the circuit current and capacitor p.d. at 10 s intervals, for the first 60 s.
4   Continue recording the current and voltage readings, at 20 s intervals, for a further 4 minutes. Reset the stopwatch to zero. Reverse the connections to the ammeter.
5   Move the switch back to position '2', and repeat the procedures of paragraphs (3) and (4) above.
6   Plot graphs of current and capacitor p.d., versus time, for both the charging and discharging cycles.
7   Submit a complete assignment report, which should include the following:

(i)   The comparison of the actual time constant (determined from the plotted graphs) to the theoretical value. Explain any discrepancy found.
(ii) Calculate several transient voltage values, and compare these with actual values plotted. Hence, justify whether or not the graphs follow an exponential change.
(iii) Explain why both the charging and discharging currents tend to 'level off' at some small value, rather than continuing to decrease to zero.

Assignment 2

To observe the affect of circuit time constant on alternating waveforms.

Apparatus:

$1 \times$ sine/squarewave signal generator
$1 \times$ variable capacitor
$1 \times$ variable resistor
$1 \times$ double-beam oscilloscope

Method:

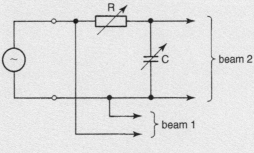

Fig. 8.41

1   Connect the circuit of Fig. 8.41, with $R = 100\,k\Omega$ and $C = 0.1\,nF$. Set the signal generator to provide a squarewave output of amplitude 5V, at a frequency of 1 kHz.
2   Observe the two waveforms displayed on the oscilloscope, and make a note of the amplitudes, time intervals and shapes of both.
3   Select $C = 1\,nF$, and repeat the above procedure.
4   Select $C = 0.1\,\mu F$, and repeat the measurements.
5   Switch the signal generator to a sinewave output, and repeat the procedures of paragraphs (2) to (4) above.
6   Transfer the connections for beam 2 of the oscilloscope so as to monitor the voltage across the resistor.
7   Set $C = 0.01\,\mu F$, $R = 5\,k\Omega$, and the signal generator to squarewave output.
8   Repeat the procedures of paragraphs (2) to (5) above.
9   Submit an assignment report, which should include scaled sketches of all the waveforms measured. Comment on the effects of different time constants on the waveforms. Identify the mathematical functions that may be performed in each case.

Chapter 1

1.1 (a) 6.772 kΩ
  (b) 1.693 kΩ
  (c) 67.73 Ω
  (d) 1.13 Ω

1.2 (a) 125.7 Ω
  (b) 1.885 kΩ
  (c) 12.57 kΩ
  (d) 3.14 MΩ

1.3 3.98 mH

1.4 26.5 μF

1.5 0.531 H

1.6 1.06 μF

1.7 238.7 Hz

1.8 15.92 kHz

1.9 (a) 42.86 mA; 1.029 W
  (b) 0.477 A; zero
  (c) 13.27 μA; zero

1.10 25 Ω

1.11 (a) 130.52 Ω
  (b) 0.92 A
  (c) +54.93 deg.; 0.5746
  (d) 63.44 W

1.12 (a) 15 V
  (b) +53.13 deg.

1.13 (a) 3 A
  (b) 0.424 mH
  (c) 27 W

1.14 (a) 5.33 Ω
  (b) 0.024 H

1.15 40.3 V

1.16 (a) 130 Ω
  (b) 1.385 A
  (c) 0.577
  (d) 143.9 W

1.17 (a) 152.3 V
  (b) 37.5 mA
  (c) 49.74 Hz
  (d) 3.733 kΩ

1.18 (a) 2.46 A
  (b) 96.13 V; 83.46 V
  (c) 236.5 W
  (d) 0.755

1.19 (a) 55.18 Ω
  (b) 4.35 A
  (c) −80.6 deg.
  (d) 39.14 V; 109.3 V; 346 V

1.20 (a) 2.85 A
  (b) 187.9 V
  (c) 412.4 V
  (d) −76.3 deg.

1.21 75.87 Hz; 12 A

1.22 0.633 H

1.23 11.25 kW

1.24 66.67 kVA; 53.33 kVAr

1.25 (a) 8.33 Ω
  (b) 5.56 Ω
  (c) 6.21 Ω
  (d) 12.8 μF
  (e) 0.677
  (f) −48.2 deg.

1.26 (a) 1.44 kW; 2.4 kVA; 1.92 kVAr
  (b) 14.4 Ω; 0.061 H

Chapter 2

2.1 (a) 1.2 A; 1.6 A
  (b) 2 A
  (c) 60 Ω

2.2 (a) 2.5 A; 3.183 A
  (b) 4.05 A
  (c) 12.35 Ω
  (d) 125 W

2.3 (a) 5 A; 8.66 A
  (b) 91.89 mH
  (c) 0.5; 60°

2.4 (a) 0.667 A; 0.942 A
  (b) 1.54 A
  (c) 8.66 Ω
  (d) 6.67 W
  (e) 13.85 VA; 12.14 VAr

2.5 7.96 μF; 100 Ω

263

2.6 (a) 0.531 A; 0.628 A
(b) 0.097 A
(c) 513.8 Ω
(d) 0 W

2.7 (a) 2.23 A
(b) 1.18 A
(c) 1.284 A
(d) −39.45°; 194.7 Ω
(e) 248 W
(f) 321.2 VA; 301.4 VAr

2.8 (a) 15 μF
(b) 1.19 A
(c) 0.457 A
(d) −21.67°
(e) 84.9 W

2.9 (a) 18.96 kHz
(b) 59.6
(c) 11.9 V
(d) 6.67 mA

2.10 (a) 2.03 μF
(b) 0.4 A
(c) 62.84 V
(d) 63.13 V

2.11 (a) 100 Ω; 22.5 mH
(b) 2.12
(c) 707 Hz

2.12 (a) 711 Hz
(b) 595.5 Hz
(c) 17.9; 1.83

2.13 (a) 1.46 kHz
(b) 25 kΩ
(c) 0.48 A
(d) 7.91

2.14 (a) 0.833 μF
(b) 1.11 mA
(c) 60
(d) 33.3 Hz

2.15 (a) 34.72 A
(b) 368.4 μF
(c) 20.83 A

2.16 (a) 234.6 μF
(b) 4.25 kVAr
(c) 23.14 A

2.17 (a) 24.45 A; 0.9425
(b) 5.33 μF

Chapter 3

3.1 3.1 A

3.2 5.37 A; 9.3 A

3.3 (b) 520 V

3.4 (a) 70.7 A
(b) 122.5 A
(c) 52.5 kW

3.5 (a) 40.9 A
(b) 40.9 A
(c) 17.55 kW

3.6 11.72 Ω; 0.1 H

3.7 (a) 51.15 A
(b) 0.866
(c) 31.4 kW

3.8 (a) 90.56 A
(b) 52.3 A

3.9 (a) 346.4 V
(b) 7.34 A
(c) 4.24 A
(d) 40.44 kW; 0.53

3.10 (a) 135 Ω; 0.207 H
(b) 3.11 kW

3.11 (a) 1.11 kW
(b) 0.736

3.12 (a) 8.2 kW
(b) −70°; 0.342

3.13 (a) 180 W
(b) 0.866

3.14 2.17 kW; 13.44 kW

3.15 (a) 19.5 A
(b) 30.14 kW

Chapter 4

4.1 1.314 A

4.2 0.116 A from B to A

4.3 5.11 V, A positive

4.4 0.635 A

4.5 1.314 A

4.6 0.116 A from B to A

4.7 1.2 V

4.8 (a) $E_0 = 13.24V; R_o = 4.71\,\Omega$

(b) $I_{sc} = 1.25\,A; R_o = 4.71\,\Omega$

(c) $4.71\,\Omega$

4.9 8.45V; 2.17 A

4.10 (a) 5.19V (B positive); 1.037 A (B to A)

(b) $0.4\,\Omega$

(c) 16.8W

4.11 2:1

Chapter 5

5.1 75

5.2 75

5.3 0.792 A

5.4 (a) 1800 V

(b) 33.75 A

(c) 60.75 kW

5.5 1.2 mWb

5.6 1.2 kV; 4 kV

5.7 (a) 0.182

(b) 97.1%

(c) 480 W

5.8 (a) 95.24%

(b) 149.4 W

(c) 95.62%

5.9 (a) 206

(b) 500 rev/min

5.10 (a) 6 poles

(b) 9.38 mWb

5.11 (a) 1332 V

(b) 2307 V

(c) 31.15 kW

5.12 (a) 14.525 kW

(b) 1.825 kW

Chapter 6

6.1 16.7 mWb

6.2 (a) 1020 rev/min

(b) 340 rev/min

6.3 415.6 V

6.4 (a) $0.2\,\Omega$

(b) 500 W

6.5 (a) 280.75V

(b) 83.3%

6.6 (a) 5 kW

(b) 280V

6.7 632 rev/min

6.8 532 rev/min

6.9 (a) 14.57 mWb

(b) 92.74 Nm

6.10 (a) 11.79 kW

(b) 84.86%

(c) 42.82 A

6.11 (a) 4.08 kW

(b) 85%

6.12 91.5%

6.13 (a) 54 kW; 90.5%

(b) 154.92 A

Chapter 8

8.1 (a) 0.183 s

(b) 6.15 mA; 0 A

(c) 0.915 s

8.2 (a) 333.3 A/s

(b) 1 A

(c) 15 ms

8.3 (a) 10.1 mA

(b) 3.25 V

(c) 22.81 V

8.4 73V; 1.03 mA

8.5 (a) 77.1 V

(b) 15.45 s

8.6 (a) $2.5\,\Omega$

(b) 9 A

8.7 (a) $100\,\Omega$

(b) 1 mA

(c) 9.49 s

(d) 90.7 μA

8.8 (a) 4 min

(b) 1.8 h

8.9   0.125 A

8.10   123.32 kΩ

8.11   (a) 416.7 A/s
       (b) 5.5 ms
       (c) 1.216 A

8.12   (a) 0
       (b) 110 V
       (c) 48.35 V

8.13   (a) 30 H
       (b) 0.547 s
       (c) 3.02 A

8.14   (b) 0.881 C
       (c) 6.95 V

8.15   100 Ω

4/9/20(

Printed and bound by CPI Group (UK) Ltd, Croydon, CR0 4YY

21/10/2024

01777097-0019